Principles of Discontinuous Dynamical Systems

Principles of Discontinuous Dynamical Systems

Marat Akhmet
Middle East Technical University, Ankara, Turkey

Springer

Marat Akhmet
Middle East Technical University
Department of Mathematics
06531 Ankara, Turkey
Marat@metu.edu.tr

ISBN 978-1-4899-9956-6 ISBN 978-1-4419-6581-3 (eBook)
DOI 10.1007/978-1-4419-6581-3
Springer New York Dordrecht Heidelberg London

Mathematics Subject Classification (2010): 34-01, 34-02, 34A12, 34A37, 34C23, 34C25, 34C28, 34D09, 34D20, 34D45, 34E05, 37G15, 37D45, 37D99, 37N05

Printed on acid-free paper

Springer is part of Springer Science+Business Media (www.springer.com)

To my beloved mother, wife and daughters:
Zhanar and Laila

Preface

The main subject of this book is discontinuous dynamical systems. These have played an extremely important role theoretically, as well as in applications, for the last several decades. Still, the theory of these systems seems very far from being complete, and there is still much to do to make the application of the theory more effective. This is especially true of equations with trajectories discontinuous at moments that are not prescribed.

The book is written not only on the basis of research experience but also, importantly, on the basis of the experience of teaching the course of Impulsive differential equations for about 10 years to the graduate students of mathematics. It is useful for a beginner as we try not to avoid any difficult instants in delivering the material. Delicate questions that are usually ignored in a research monograph are thoroughly addressed. The standard material on equations with fixed moments of impulses is presented in a compact and definitive form. It contains a large number of exercises, examples, and figures, which will aid the reader in understanding the enigmatic world of discontinuous dynamics. The following peculiarity is very important: the material is built on the basis of close parallelism with ordinary differential equations theory. For example, even higher order differentiability of solutions, which has never been considered before, is presented with a full definition and detailed proofs. At the same time, the definition of the derivatives as coefficients of the expansion is fruitfully used, which is very rare in the theory of ordinary differential equations. Moreover, the description of stability, continuous and differentiable dependence of solutions on initial conditions, and right-hand side, chaotic ingredients is given on a more strong functional basis than that of ordinary differential equations.

The book is attractive to an advanced researcher, since a strong background for the future analysis of all theoretical and application problems is built. It will benefit scientists working in other fields of differential equations with discontinuities of various types, since it reflects the experience of the author in working on these subjects. We would like to emphasize that the basics of discontinuous flows are for the first time rigorously laid out so that all the attributes of dynamical systems are present. Hence, there is plenty of room for extending all the results of continuous, smooth and analytic dynamics to the systems with discontinuities.

The content of the book is a good background for the application in vibro-mechanisms theory, mechanisms with friction, biology, molecular biology, physiology, pharmacology, secure communications, neural networks, and other real world problems involving discontinuities.

Chapters 5–10 contain the core research contributions. Chapters 1–4 present preliminaries for the theory and elements of differential equations with fixed moments of discontinuity. Chapters 1–8 provide sufficient material for a standard one-semester graduate course. It is natural to finalize a general theory with more specific results. For this reason, in the last two chapters (9 and 10) we discuss Hopf bifurcation of periodic discontinuous solutions, Devaney's chaos, and the Shadowing property for discontinuous dynamical systems.

The author expresses his gratitude to his students who have contributed to the preparation of this book: Duygu Arugaslan, Cemil Buyukadali, Mehmet Turan, and Enes Yilmaz.

Contents

Chapter 1
Introduction

Nowadays, many mathematicians agree that discontinuity as well as continuity should be considered when one seeks to describe the real world more adequately. The idea that, besides continuity, discontinuity is a property of motion is as old as the idea of motion itself. This understanding was strong in ancient Greece. For example, it was expressed in paradoxes of Zeno. Invention of calculus by Newton and Leibniz in its last form, and the development of the analysis adjunct to celestial mechanics, which was stimulating for the founders of the theory of dynamical systems, took us away from the concept of discontinuity. The domination of continuous dynamics, and also smooth dynamics, has been apparent for a long time. However, the application of differential equations in mechanics, electronics, biology, neural networks, medicine, and social sciences often necessitates the introduction of discontinuity, as either abrupt interruptions of an elsewhere continuous process (impulsive differential equations) or in the form of discrete time setting (difference equations). If difference equations may be considered as an instrument of investigation of continuous motion through, for example, Poincaré maps, impulsive differential equations seem appropriate for modeling motions where continuous changes are mixed with impact type changes in equal proportion. Recently, it is becoming clear that to discuss real world systems that (1) exist for a long period of time, or (2) are multidimensional, with a large number of dependent variables, researchers resort to differential equations with: (1) discontinuous trajectories (impulsive differential equations); (2) switching in the right-hand side (differential equations with discontinuous right-hand side); (3) some coordinates ruled by discrete equations (hybrid systems); (4) disconnected domains of existence of solutions (time scale differential equations), where these properties may be combined in a single model.

The theory of equations with discontinuous trajectories has been developed through applications [14, 16, 38, 41, 43, 50, 52, 53, 56, 57, 70, 71, 75, 79, 89, 91, 99, 101, 103, 107–109, 115, 121, 123, 125–127, 130, 144, 145, 155, 158–160, 162] and theoretical challenges [4–9, 19, 32–36, 65, 69, 75, 85, 95–97, 99–101, 103, 110, 111, 118–124, 135–142, 151–153].

We give a limited number of references, since this work was written as a textbook rather than a research monograph, and secondly, sources related to systems with nonfixed moments of discontinuity were preferentially presented.

M. Akhmet, *Principles of Discontinuous Dynamical Systems*,
DOI 10.1007/978-1-4419-6581-3_1, © Springer Science+Business Media, LLC 2010

Our main objective is to present the theory of differential equations with solutions that have discontinuities either at the moments when the integral curves reach certain surfaces in the extended phase space (t, x), as time t increases (decreases), or at the moments when the trajectories enter certain sets in the phase space x. That is, the moments when the solutions have discontinuities are not prescribed. Notably, the systems with nonprescribed times of discontinuity were first introduced in [91, 123, 124], manuscripts in applied mathematics, which underscores the practical importance of the theory of equations with variable moments of impulses. Differential equations with fixed moments of impulses were the next to be studied. These serve as an auxiliary instrument for the study of the above named systems in the same way as nonautonomous equations play a role in the analysis of autonomous systems through linearization. For that reason, we provide a more extensive discussion of the theory of equations with fixed moments of impulses, than might otherwise seem necessary. It takes the first four chapters of the manuscript. We thoroughly describe the solutions of these equations, consider the existence and uniqueness of solutions and their dependence on parameters. The problem of extension of a solution for both increasing and decreasing time is investigated. For example, we prove the Gronwall–Bellman Lemma for piecewise continuous functions, and the integral representation formulas, for decreasing time, as well as for increasing time. This extension of the results is obviously necessary to explore dynamical systems' properties in the fullest form, as required for applications. Since the moments of discontinuities are different for different solutions, the equations are nonlinear. Equations with nonfixed moments of discontinuity create a great number of opportunities for theoretical inquiry, as well as theoretical challenges. This is due to the structure of these systems, namely the three components: a differential equation, an impulsive action, and the surfaces of discontinuity which are involved in the process of governing the motion. Therefore, in addition to the features of the ordinary differential equations, we may vary the properties of the maps, which transform the phase point at the moment of impulse, and try various topological and differential characteristics of the surfaces of discontinuity to produce one or another interesting theoretical phenomenon, or satisfy a desired application property.

Effective methods of investigation of systems with nonfixed moments of impulsive action can be found in [2–4, 32–36, 65, 69, 95–97, 123, 124, 135, 136, 142, 152, 153]. Theoretical problems of nonsmooth dynamics and discontinuous maps [17, 19, 38, 48, 51, 66, 68, 86, 92, 93, 146] are also close to the subject matter of our book.

The present book plays its own modest role in attracting the attention of scientists, first of all mathematicians, to the symbiosis of continuity and discontinuity in the description of a motion.

The book presented to the attention of the reader is to be viewed first and foremost as a textbook on the theory of discontinuous dynamical systems. There is some similarity between the content of this book and that of the monographs on ordinary differential equations. Accordingly, we deliver some standard topics: description of the systems, definition of solutions, local existence and uniqueness theorems, extension of solutions, dependence of solutions on parameters. It is our conviction that many results of the theory of equations with impulses (if not all), that at the moment appear as very specific, in fact, have their counterparts in the theory of ordinary differential equations. We take up the task of extending the parallels with the theory of continuous (smooth) dynamical systems. It seems appropriate to place the results on the existence of periodic solutions and Hopf bifurcation of periodic solutions in the final part of the book. The last chapter is devoted to complex motions, in whose description we use ingredients of Devaney chaos. It is noteworthy that the method of creation of chaos through impulses does not have analogs in continuous dynamics yet. We bring up only a few examples to illustrate the possibilities for application.

We use a powerful analytical tool of B-equivalence, which was introduced and developed in our papers. The method was created especially for the investigation of systems with solutions that have discontinuities at variable moments of time [1–4, 25–37]. But it can also be applied to differential equations with discontinuous right-hand-side [13, 15, 27, 34] and differential equations on variable time scales with transition conditions [20]. The method is effective in the analysis of chaotic systems [8–11], as well.

In the last decades, the exceptional role of differential equations with impulses at variable times in dealing with problems of mechanisms with vibrations has been perceived. Collision-bifurcations, oscillations, and chaotic processes in this mechanisms have been investigated in many papers and books [67, 79, 118, 119, 144, 151, 156]. We are very confident that the content of this book will give a strong push to the development of this field, as well as other related areas of research, where a discontinuity appears.

Let us consider the following examples, which highlight the modeling role of discontinuous dynamics.

Example 1.1. Consider a mechanical model consisting of a bead B bouncing on a massive, sinusoidally vibrating table P (see Fig. 1.1). Such a system has been investigated in [71, 79, 89, 133, 158]. We assume that the table is so massive that

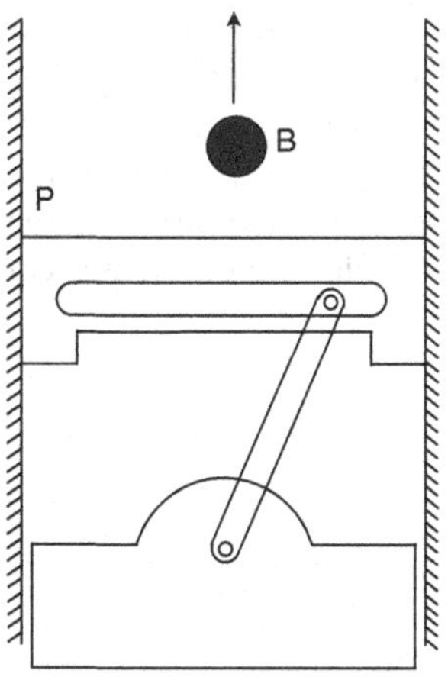

Fig. 1.1 A model consisting of a bead bouncing on a vibrating table

it does not react to collisions with the bead and moves according to the law $X = X_0 \sin \omega t$. The motion of the bead between collisions is given by the formula

$$x = \frac{-g(t - t_0)^2}{2} + x_0'(t - \phi) + x_0, \tag{1.1}$$

where x_0 and x_0' are, respectively, the values of the coordinate and the velocity of the bead at the instant $t = \phi$ immediately after collision, and $g = 9.8\, m/s^2$ is the gravitational constant. The change of the velocity of the bead at the moment of the hit is given by the following relation:

$$R = \frac{X_+' - x_+'}{x_-' - X_-'}. \tag{1.2}$$

Here R is the restitution coefficient ($0 < R \leq 1$), X_-', x_-', X_+', and x_+' are the velocities of the table and the bead before and after the strike, repectively, $(X_+' = X_-')$.

Among the results of investigation of the model, one can mention those in [71], where the period-doubling bifurcation, as well as chaos emergence, is discussed. If we write $x_1 = x, x_2 = x', \tau_i(x_1) = \arcsin(x_1/X_0) + (\pi/\omega)i$, where i are integers, then, using (1.1) and (1.2), one can construct a suitable mathematical model in the form of the following nonlinear system of differential equations with impulsive actions:

$$\begin{aligned} x_1' &= x_2, \\ x_2' &= -g, \\ \Delta x_2|_{t=\tau_i(x_1)} &= (1 + R)[x_0\omega \cos(\omega\tau_i(x_1)) - x_2]. \end{aligned} \tag{1.3}$$

Example 1.2. Consider a mechanical model of the oscillator consisting of a cart C (see Fig. 1.2), which can impact against a rigid wall W, and is subjected to an external force $H \sin(\omega t + \gamma)$. There is an elastic element S. The wall is at the distance B from the origin of the coordinate system, which is placed at the equilibrium point. The change of velocity of the cart at the moment of the hit against the wall is given by the relation $x_+' = -Rx_-'$, where R is the restitution coefficient ($0 < R \leq 1$), and x_-' and x_+' are the velocities of the cart before and after the strike, respectively. One can easily find a mathematical model of the system, which takes the form of the following differential equations with impulses:

$$\begin{aligned} x_1' &= x_2, \\ x_2' &= -cx_1 + H \sin(\omega t + \gamma), \\ \Delta x_2|_{x_1=-B} &= (1 + R)x_2, \end{aligned} \tag{1.4}$$

where $x_1 = x, x_2 = x'$.

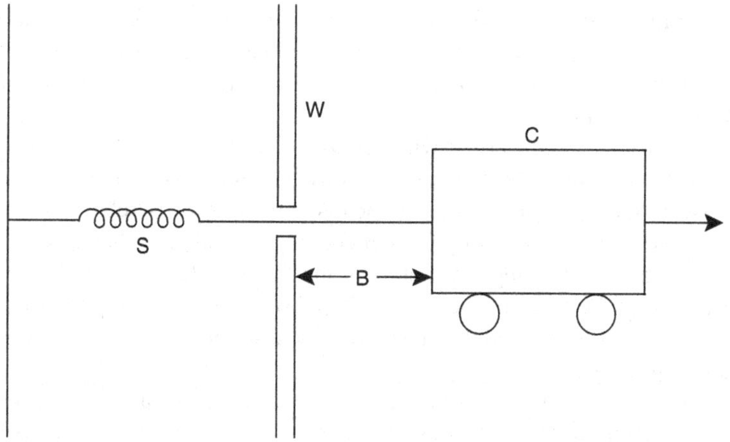

Fig. 1.2 A model of an oscillator consisting of a cart C, which can impact against a rigid wall W

Systems (1.3) and (1.4) are typical examples of differential equations with variable moments of impulses discussed in the book. The first system has solutions that exhibit discontinuity when they reach surfaces in the extended phase space. Solutions of the other system have jumps at the moments when they cross a set in the phase space of the equation.

The book is organized as follows:

We start with the description of differential equations with fixed moments of impulses in the second chapter. The characteristics of the sets of discontinuity moments are listed, and the spaces of piecewise continuous functions are introduced. The extension of solutions is presented in a very detailed manner. The theorems on local and global existence, and uniqueness of solutions are proved. The continuous dependence of solutions on initial conditions and the right-hand side are discussed.

The third chapter is devoted to the generalities of stability and periodic solutions of differential equations with fixed moments of impulses. Definitions of stability, the description of periodic systems, and illustrating examples are provided.

The basics of linear impulsive systems are the focus of the fourth chapter: Linear homogeneous systems; Linear nonhomogeneous systems; Linear periodic systems; Spaces of solutions; Stability of linear systems.

The next, fifth chapter is one of the main parts of the book. Nonautonomous differential equations with impulses at variable moments of time, whose solutions have jumps at the moments of intersection with surfaces in the extended phase space, are considered. In this chapter, we provide all conditions that make the investigation of these equations convenient. Namely, the conditions that guarantee the absence of beating of the solutions against the surfaces of discontinuity, and the conditions that preserve the ordering of the intersection with the surfaces. Moreover, we obtain conditions that allow the reduction to equations with fixed moments of impulses. It should be emphasized that the results concerning the dependence of solutions on the initial conditions and on the right-side, and stability are presented in the full form for the first time in the literature. Also, for the first time conditions for the extension of solutions to the left are formulated. The main auxiliary concepts are: the topology in the set of discontinuous functions. A general nonlinear case is considered, and quasilinear systems are investigated.

The sixth chapter is concerned with differentiability properties of solutions of nonautonomous differential equations with variable moments of impulses with respect to the initial conditions and parameters. The subject is relatively new for discontinuous dynamics, especially for higher order derivatives. What makes this investigation possible is the implementation of the B-equivalence method. The same can be said about the issue of the analyticity of solutions. We propose a uniform approach so that not only solutions themselves but also their discontinuity moments can be differentiated.

The results on smoothness from the previous chapter are used in the seventh chapter to develop the method of small parameter for quasilinear systems. Both critical and noncritical cases for the existence of periodic solutions are discussed. Practically useful algorithms are derived.

Chapter 8 is the central part of the book. We obtain conditions sufficient to shape a motion that is very similar to the flow of an autonomous ordinary differential equation so that all the properties of the dynamical system – extension of all solutions on $\mathbb{R}$, continuous dependence on the initial value, the group property and uniqueness – are preserved. Differentiability in the initial value is considered. In fact, B-smooth discontinuous flows are obtained.

The last two chapters revolve around more specific topics. The ninth chapter develops the mechanisms for discovering the Hopf bifurcation of a discontinuous dynamical system. Additionally, the question of the persistence of focus and the problems of distinguishing the focus and the center in the critical case are discussed as preliminaries.

We consider complex behavior of a discontinuous dynamics in the tenth chapter. For a special initial value problem, where moments of impulses are generated through map iterations, analogs of all Devaney's ingredients, as well as the shadowing property, are studied. Examples illustrating the existence of the chaotic attractor and the intermittency phenomenon are provided.

Chapter 2
Description of the System with Fixed Moments of Impulses and Its Solutions

2.1 Spaces of Piecewise Continuous Functions

Let $\mathbb{R}$, $\mathbb{N}$, and $\mathbb{Z}$ be the sets of all real numbers, natural numbers, and integers, respectively. Denote by $\theta = \{\theta_i\}$ a strictly increasing sequence of real numbers such that the set $\mathcal{A}$ of indexes i is an interval in $\mathbb{Z}$.

Definition 2.1.1. θ is a B-sequence, if one of the following alternatives is valid:

(a) $\theta = \emptyset$;
(b) θ is a nonempty and finite set;
(c) θ is an infinite set such that $|\theta_i| \to \infty$ as $|i| \to \infty$.

Example 2.1.1. θ with $\mathcal{A} = \{-1, 1, 2\}$, and $\theta_{-1} = -5, \theta_1 = \pi, \theta_2 = 7$, satisfies condition (b).

Example 2.1.2. $\theta_i = i + \frac{1}{3}$, where $i \geq -102$, is a B-sequence of type (c).

Example 2.1.3. $\theta_i = -i + \frac{1}{5}$, where $i \in \mathbb{Z}$, is not a B-sequence.

It is obvious that any B-sequence has no finite limit points.

Example 2.1.4. $\theta_i = 1 - \frac{1}{i+3}$, where $i = 1, 2, 3, \ldots$, is not a B-sequence.

Denote by Θ the union of all B-sequences.

Fix a sequence $\theta \in \Theta$.

Definition 2.1.2. A function $\phi : \mathbb{R} \to \mathbb{R}^n, n \in \mathbb{N}$, is from the set $\mathcal{PC}(\mathbb{R}, \theta)$ if :

(i) it is left continuous;
(ii) it is continuous, except, possibly, points of θ, where it has discontinuities of the first kind.

The last definition means that if $\phi(t) \in \mathcal{PC}(\mathbb{R}, \theta)$, then the right limit $\phi(\theta_i+) = \lim_{t\to\theta_i+} \phi(t)$ exists and $\phi(\theta_i-) = \phi(\theta_i)$, where $\phi(\theta_i-) = \lim_{t\to\theta_i-} \phi(t)$, for each $\theta_i \in \theta$.

Let $\mathcal{PC}(\mathbb{R}) = \cup_{\theta\in\Theta}\mathcal{PC}(\mathbb{R}, \theta)$.

M. Akhmet, *Principles of Discontinuous Dynamical Systems*,
DOI 10.1007/978-1-4419-6581-3_2, © Springer Science+Business Media, LLC 2010

Definition 2.1.3. A function $\phi : \mathbb{R} \to \mathbb{R}^n$ is from the set $\mathcal{PC}^1(\mathbb{R}, \theta)$ if $\phi(t), \phi'(t) \in \mathcal{PC}(\mathbb{R}, \theta)$, where the derivative at points of θ is assumed to be the left derivative.

In what follows, in this section, $T \subset \mathbb{R}$ is an interval in $\mathbb{R}$. For simplicity of notation, θ is not necessary a subset of T.

Definition 2.1.4. A function $\phi : T \to \mathbb{R}^n$ is from the set $\mathcal{PC}_r(T, \theta)$ if it has a continuation from $\mathcal{PC}(\mathbb{R}, \theta)$.

Definition 2.1.5. A function $\phi : T \to \mathbb{R}^n$ is from the set $\mathcal{PC}_r^1(T, \theta)$ if it has a continuation from $\mathcal{PC}^1(\mathbb{R}, \theta)$.

Exercise 2.1.1. Prove that Definitions 2.1.4 and 2.1.5 are equivalent to the following two Definitions 2.1.6 and 2.1.7, respectively.

Definition 2.1.6. A function $\phi : T \to \mathbb{R}^n$ is from the set $\mathcal{PC}_r(T, \theta)$ if:

(i) ϕ is left continuous;
(ii) ϕ is continuous, except, possibly, points of θ, where it has discontinuities of the first kind;
(iii) if an end point of T is finite, then there exists the one-sided limit of ϕ.

Definition 2.1.7. A function $\phi : T \to \mathbb{R}^n$ belongs to the set $\mathcal{PC}_r^1(T, \theta)$ if:

(i) $\phi \in \mathcal{PC}_r(T, \theta)$;
(ii) $\phi'(t) \in \mathcal{PC}_r(T, \theta)$, where the derivative at finite end points is a one-sided derivative and the derivative at points of θ is assumed to be the left derivative.

We shall use also the following definitions.

Definition 2.1.8. A function $\phi : T \to \mathbb{R}^n$ is from the set $\mathcal{PC}(T, \theta)$ if:

(i) ϕ is left continuous;
(ii) ϕ is continuous, except, possibly, points from θ, where it has discontinuities of the first kind.

Definition 2.1.9. A function $\phi : T \to \mathbb{R}^n$ belongs to the set $\mathcal{PC}^1(T, \theta)$ if:

(i) $\phi \in \mathcal{PC}(T, \theta)$;
(ii) $\phi'(t) \in \mathcal{PC}(T, \theta)$, where the derivative at finite end points is assumed to be a one-sided derivative, and the derivative at points of θ is assumed to be the left derivative.

Exercise 2.1.2. Assume that an interval T is one of the following types: the axis $\mathbb{R}$; a section $[a, b]$; a half-axis $[a, \infty)$; a half-axis $(-\infty, b], a, b \in \mathbb{R}$. Prove that $\mathcal{PC}_r(T, \theta) = \mathcal{PC}(T, \theta)$. Otherwise, $\mathcal{PC}_r(T, \theta) \subset \mathcal{PC}(T, \theta)$.

Exercise 2.1.3. Does the greatest integer function belong to $\mathcal{PC}(\mathbb{R})$?

2.2 Description of the System

Let $I \subseteq \mathbb{R}$ be an open interval, θ a nonempty B-sequence with set of indexes $\mathcal{A}$, and $G \subseteq \mathbb{R}^n, n \in \mathbb{N}$, an open connected set. Consider a function $f : I \times G \to \mathbb{R}^n$ and a map $J : \mathcal{A} \times G \to \mathbb{R}^n$, which we shall write as $f(t, x)$ and $J_i(x)$, respectively. Assume throughout this chapter that f is a continuous function.

For a fixed $i \in \mathcal{A}$, we introduce a *transition operator* $\Pi_i x \equiv x + J_i(x)$ on G.

Let $x(t)$ be a function defined in a neighborhood of a number $\xi \in \mathbb{R}$. We set $\Delta x|_{t=\xi} \equiv x(\xi+) - x(\xi)$, assuming that the limit $x(\xi+) = \lim_{t\to\xi+} x(t)$ exists. Applying the transition operator Π_i, define the following *equation of jumps*

$$\Delta x|_{t=\theta_i} = \Pi_i x(\theta_i) - x(\theta_i)$$

or

$$\Delta x|_{t=\theta_i} = J_i(x(\theta_i)). \tag{2.1}$$

Definition 2.2.1. We call the pair, which consists of the equation of jumps (2.1) and the ordinary differential equation

$$x' = f(t, x), \tag{2.2}$$

where the derivative at points of θ is assumed to be the left derivative, a discontinuous vector field.

Thus, the discontinuous vector field has the form

$$\begin{aligned} x' &= f(t, x), \\ \Delta x|_{t=\theta_i} &= J_i(x). \end{aligned} \tag{2.3}$$

We shall call this field, (2.3), an *impulsive differential equation* . The domain of the equation is the set $\Omega = I \times \mathcal{A} \times G$.

In our book we make the following assumption, which is valid everywhere unless otherwise stated.

(M0) Each solution $\phi(t), \phi(t_0) = x_0, (t_0, x_0) \in I \times G$, of ordinary differential equation (2.2) exists and is unique on any interval of existence. It has an open maximal interval of existence, and any limit point of the set $(t, \phi(t))$, as t tends to an end point of the interval, is a boundary point of $I \times G$.

Let us remember that (M0) is valid if, for example, $f(t, x)$ satisfies a local Lipschitz condition.

Definition 2.2.2. [60] The function $f(t, x)$ satisfies a local Lipschitz condition, with respect to x, in $I \times G$ if for any compact subset K of $I \times G$ there exists a positive number L_K, such that

$$\|f(t, x) - f(t, y)\| \le L_K \|x - y\|$$

for any $(t, x), (t, y) \in K$.

2.3 Description of Solutions

Assume that $\phi : T \to \mathbb{R}^n$, where $T \subseteq I$ is an open interval, is a solution of (2.3). That is, it satisfies differential equation (2.2) and equation of jumps (2.1).

Theorem 2.3.1. *The solution ϕ belongs to $\mathcal{PC}^1(T, \theta)$.*

Proof. Indeed, by the differentiability of ϕ, it is continuous on $T\backslash\theta$. Consequently, as f is continuous, we have that ϕ' is continuous on the same set. The left differentiability implies that ϕ is a left-continuous function at the points from θ. Hence, ϕ is a left-continuous function on T. Now, ϕ' is a left-continuous function at all points of θ, since of the continuity of f. Further, the formula

$$\Delta\phi|_{t=\theta_i} = J_i(\phi(\theta_i))$$

yields that the right limit of the solution ϕ exists at the points of θ and

$$\phi(\theta_i+) = \phi(\theta_i) + J_i(\phi(\theta_i)).$$

Using the last equality in (2.2), one can obtain that the limit

$$\lim_{t\to\theta_i+} \phi'(t) = \phi'(\theta_i+)$$

exists. Thus, we can conclude that $\phi \in \mathcal{PC}^1(T, \theta)$. The theorem is proved. □

Exercise 2.3.1. Assume that a function $\phi : T \to \mathbb{R}^n$, where $T \subset I$ is a finite closed interval, is a solution of equation (2.3). Prove that $\phi \in \mathcal{PC}^1_r(T, \theta)$.

Example 2.3.1. Consider the following system:

$$\begin{aligned} x' &= 0, \\ \Delta x|_{t=i} &= (-1)^i, \end{aligned} \tag{2.4}$$

where $t, x \in \mathbb{R}$, and the function

$$\phi(t) = \begin{cases} 1, & \text{if } 2i < t \leq 2i + 1, \\ 0, & \text{if } 2i - 1 < t \leq 2i, i \in \mathbb{Z}. \end{cases} \tag{2.5}$$

It is easy to check that ϕ satisfies the differential equation. Moreover, $\Delta\phi|_{t=2k} = \phi(2k+) - \phi(2k) = 1 - 0 = 1$ and $\Delta\phi|_{t=2k-1} = 0 - 1 = -1$. That is, ϕ is a solution of (2.4). One can see that $\phi \in \mathcal{PC}^1(\mathbb{R}, \theta)$, if $\theta_i = i, i \in \mathbb{Z}$.

Exercise 2.3.2. Prove that the solution ϕ of (2.4) is a 2-periodic function.

In what follows, we consider existence and extension of a solution of the impulsive differential equation (2.3). Because of the singularity at moments of discontinuity, the procedure of extension is more complex than that of ordinary differential equations.

We shall use the following definitions, which are similar to those of ordinary differential equations [60, 77, 98].

Definition 2.3.1. A solution $x_1(t) : T_1 \to G$ of (2.3) is said to be a continuation of a solution $x(t) : T \to G$ on T_1, and $x(t)$ is said to be continuable on T_1, if $T \subset T_1$ and $x(t) = x_1(t)$ on T.

Definition 2.3.2. An interval T is called a right maximal interval of existence of a solution $x(t)$ of (2.3) if there is no continuation of the solution on an interval T_1 with right end point greater, than the end point of T.

Similarly one can give a definition of a left maximal interval of existence.

Definition 2.3.3. An interval T is said to be a maximal interval of existence of a solution $x(t)$ of (2.3) if it is both a right and left maximal interval of the solution.

Let a point $(t_0, x_0) \in I \times G$ be given. Denote by $x(t) = x(t, t_0, x_0)$ a solution of the initial value problem (2.3) and $x(t_0) = x_0$.

Extension over a maximal interval of existence. We will extend the solution, if:

(a) t is increasing, $t \geq t_0$;
(b) t is decreasing, $t \leq t_0$.

The case (a), in its own turn, consists of two sub-cases:

(a1) $t_0 \neq \theta_i, i \in \mathcal{A}$;
(a2) $t_0 = \theta_j$ for some $j \in \mathcal{A}$.

In the sequel, we denote by $\phi(t, \kappa, z), \kappa \in I, z \in G$, a solution of ordinary differential equation (2.2) such that $\phi(\kappa, \kappa, z) = z$.

Let us consider the sub-case $(a1)$. That is, the initial moment is not a discontinuity point. To be more concrete, suppose that $\theta_{j-1} < t_0 < \theta_j$, for a fixed $j \in \mathcal{A}$, (see Fig. 2.1). Denote by $[t_0, r), t_0 < r$, the right maximal interval of the solution $\phi(t, t_0, x_0)$. If $r \leq \theta_j$, then $[t_0, r)$ is the maximal interval of $x(t)$, and

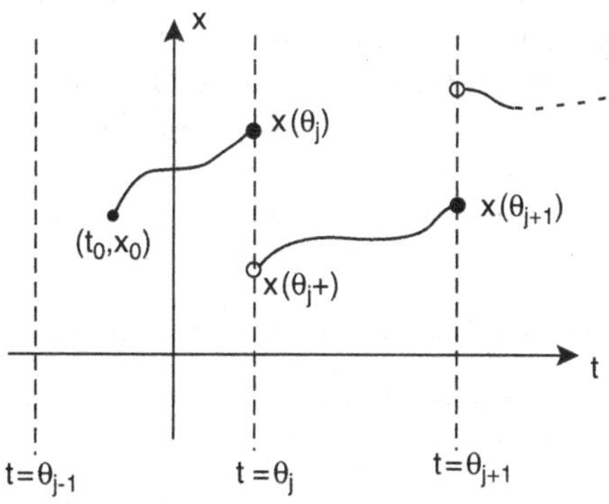

Fig. 2.1 Extension of a solution of system (2.3)

$x(t) = \phi(t, t_0, x_0)$. Otherwise, $r > \theta_j$, and if $\Pi_j x(\theta_j) \notin G$, then $x(t)$ is not continuable beyond $t = \theta_j$, and the right maximal interval of $x(t)$ is $[t_0, \theta_j]$. If $\Pi_j x(\theta_j) \in G$, then $x(t)$ can be extended to the right as $x(t) = \phi(t, \theta_j, x(\theta_j+))$, where $x(\theta_j+)) = \Pi_j x(\theta_j)$. Denote by $[\theta_j, r)$ the right maximal interval of $\phi(t, \theta_j, x(\theta_j+))$. If $r \leq \theta_{j+1}$, then $[t_0, r)$ is the right maximal interval of $x(t)$, and $x(t) = \phi(t, \theta_j, x(\theta_j+)), t \in [\theta_j, r)$. If $r > \theta_{j+1}$, and $\Pi_{j+1}x(\theta_{j+1}) \notin G$, then the maximal interval of existence of $x(t)$ is $[t_0, \theta_{j+1}]$. If $r > \theta_{j+1}$ and $\Pi_{j+1}x(\theta_{j+1}) \in G$, then $x(t)$ can be continued as $\phi(t, \theta_{j+1}, x(\theta_{j+1}+))$, and so on.

If $\mathcal{A}$ is a finite set, then by using a finite number of steps we shall define the right maximal interval of $x(t)$, bounded or unbounded. Now, assume that the set $\mathcal{A}$ is unbounded on the right. There are two possibilities of the extension of the solution: either the number of steps is infinite, and the maximal interval is unbounded on the right or the number of admissible steps is finite and the right maximal interval is definitely bounded.

Now, consider the subcase $(a2), t_0 = \theta_j$ for some $j \in \mathcal{A}$. That is, the initial moment is a point of discontinuity. If $\Pi_j x(\theta_j) \notin G$, then $x(t)$ does not exist for $t > t_0$. If $\Pi_j x(\theta_j) \in G$, then $x(t_0+) = \Pi_j x(\theta_j)$. That is, the solution jumps at t_0, and the further discussion is the same as that of sub-case $(a1)$.

Consider the left extension of the solution. That is, the case (b). It is useful to start with the following exercise.

Exercise 2.3.3. Explain why an initial moment cannot be a discontinuity point for a left extension.

Fix $j \in \mathcal{A}$ such that $\theta_j < t_0 \leq \theta_{j+1}$. Denote by $(l, t_0], l < t_0$, the left maximal interval of existence of the solution $\phi(t, t_0, x_0)$. If $l > \theta_j$, then $(l, t_0]$ is a left maximal interval of $x(t)$. If $l = \theta_j$ and the limit $\phi(\theta_j+, t_0, x_0)$ does not exist, then $(l, t_0]$ is a left maximal interval of existence of $x(t)$. Otherwise, if $l = \theta_j$ and the limit $\phi(\theta_j+, t_0, x_0)$ exists, or if $l < \theta_j$, then $x(\theta_j+) = \phi(\theta_j+, t_0, x_0)$. Now, if the equation

$$x(\theta_j+) = v + J_j(v) \tag{2.6}$$

is solvable with respect to v in G, then take the solution of this equation as $x(\theta_j)$, and continue $x(t) = \phi(t, \theta_j, x(\theta_j))$ at the right of θ_j. If (2.6) does not have a solution in G, then $(\theta_j, t_0]$ is a left maximal interval of existence of $x(t)$. Remark that (2.6) may have several solutions, and even infinitely many. We assume that one can be chosen, and we are not busy with the uniqueness problem this time. Proceeding in this way, one could finally construct a left maximal interval of $x(t)$.

Unite the left and right maximal intervals to obtain the maximal interval of $x(t)$.

Exercise 2.3.4. Solve the following six problems.

1. Complete the discussion of the left maximal interval similarly to that of the right maximal interval.

2. Explain why the solution $x(t, t_0, x_0)$ of (2.3) may have more than one maximal intervals.
3. Explain why the solution $x(t, t_0, x_0)$ of (2.3) may have more than one left maximal intervals.
4. Explain why the maximal interval of existence of the solution $\phi(t, t_0, x_0)$ of ordinary differential equation (2.2) is open.
5. Why a solution of (2.3) may have a right-closed maximal interval?
6. Why the left maximal intervals are not closed?

Exercise 2.3.5. Consider the system

$$\begin{aligned} x' &= 0, \\ \Delta x|_{t=i} &= x^{-1}, \end{aligned} \tag{2.7}$$

where $t, x \in \mathbb{R}, i \in \mathbb{Z}$. Verify that the solution $x(t, 0, 2)$ of (2.7) exists and is unbounded on $[0, \infty)$.

Hint: To prove the unboundedness, use a contradiction.

Theorem 2.3.2. *(Local existence theorem) Suppose $f(t, x)$ is continuous on $I \times G$ and $\Pi_i G \subseteq G, i \in \mathcal{A}$. Then for any $(t_0, x_0) \in I \times G$ there is $\alpha > 0$ such that a solution $x(t, t_0, x_0)$ of (2.3) exists on $(t_0 - \alpha, t_0 + \alpha)$.*

Theorem 2.3.3. *Assume that condition (M0) is valid. Then each solution of (2.3) has a maximal interval of existence and for this interval the following alternatives are possible:*

(1) it is an open interval (α, β) such that any limit point of the set $(t, x(t))$ as $t \to \alpha$ or $t \to \beta$ belongs to the boundary of $I \times G$;
(2) it is a half-open interval $(\alpha, \beta]$, where β is a member of θ and any limit point of the set $(t, x(t))$ as $t \to \alpha$ belongs to the boundary of $I \times G$.
(3) it is a half-open interval $(\alpha, \beta]$, where α and β are elements of θ, the limit $x(\alpha+)$ exists and it is an interior point of G.
(4) it is an open interval (α, β) such that any limit point of the set $(t, x(t))$ as $t \to \beta$ belongs to the boundary of $I \times G$, the limit $x(\alpha+)$ exists and it is an interior point of G.

Theorem 2.3.4. *(Uniqueness theorem) Assume that $f(t, x)$ satisfies a local Lipschitz condition, and every equation*

$$x = v + J_i(v), \tag{2.8}$$

$i \in \mathcal{A}, x \in G$, has at most one solution with respect to v. Then any solution $x(t, t_0, x_0), (t_0, x_0) \in I \times G$, of (2.3) is unique. That is, if $y(t, t_0, x_0)$ is another solution of (2.3), and the two solutions are defined at some $t \in I$, then $x(t, t_0, x_0) = y(t, t_0, x_0)$.

Exercise 2.3.6. Prove the last three theorems.

Remark 2.3.1. The results of our book can be easily expanded for more general impulsive systems, if the function f in (2.3) is discontinuous and is of the following type. Without loss of generality assume that $I = \mathbb{R}$ and $\mathcal{A} = \mathbb{Z}$. Consider $f(t,x)$ such that it is continuous in each region $(\theta_i, \theta_{i+1}] \times G, i \in \mathbb{Z}$, and there exists a continuation $f_e^i(t,x)$ of this function on the set $[\theta_i, \theta_{i+1}] \times G$ for each $i \in \mathbb{Z}$. To demonstrate how the discussion should be arranged in this case, we consider the extension of the solution $x(t) = x(t, t_0, x_0)$ for $t \geq t_0$. Fix $j \in \mathbb{Z}$, such that $\theta_{j-1} < t_0 < \theta_j$. The beginning of the extension is identical to that of with continuous f. That is, assume that $x(t) = \phi(t, t_0, x_0), t \in [t_0, \theta_j]$ where ϕ is the solution of (2.2). Let $\Pi_j x(\theta_j) \in G$. Now, $x(t)$ is continuable to the right, and $x(t) = \psi(t, \theta_j, x(\theta_j+))$, where $x(\theta_j+) = \Pi_j x(\theta_j)$, and ψ is not the solution of (2.2) as it was above, but of the system

$$x' = f_e^j(t,x), \tag{2.9}$$

where $t \in [\theta_j, \theta_{j+1}]$. Proceeding in this way one can find the right maximal interval of $x(t)$.

Impulsive systems with the discontinuous right-hand side are needed for applications and investigation methods. For example, they emerge if one linearizes an impulsive system around a certain solution, (see, for instance, Chap. 6).

2.4 Equivalent Integral Equations

In this small section we construct an integral equation, which has the common set of solutions with the original impulsive differential equation (2.3). To prove the equivalence assertion, we shall need the result of the following exercise:

Exercise 2.4.1. Let ϕ and ψ be functions from $\mathcal{PC}^1(T, \theta)$ such that:

(1) $\phi(t_0) = \psi(t_0)$ for some $t_0 \in T$;
(2) $\phi'(t) = \psi'(t)$ for all $t \in T$, (the derivative is the left derivative at points of θ, and it is an one-sided derivative at the end points of T);
(3) $\Delta\phi|_{t=\theta_i} = \Delta\psi|_{t=\theta_i}, i \in \mathcal{A}$.

Prove that $\phi(t) = \psi(t)$ for all $t \in T$.

Theorem 2.4.1. *A function $\phi(t) \in \mathcal{PC}^1(T, \theta), \phi(t_0) = x_0$, is a solution of (2.3) if and only if*

$$\phi(t) = \begin{cases} x_0 + \int_{t_0}^t f(s, \phi(s))ds + \sum_{t_0 \leq \theta_i < t} J_i(\phi(\theta_i)), & t \geq t_0, \\ x_0 + \int_{t_0}^t f(s, \phi(s))ds - \sum_{t \leq \theta_i < t_0} J_i(\phi(\theta_i)), & t < t_0. \end{cases} \tag{2.10}$$

Proof. Necessity. Let $\phi(t)$ be a solution of (2.3) on T. Define a function

$$\psi(t) = \begin{cases} x_0 + \int_{t_0}^t f(s, \phi(s))ds + \sum_{t_0 \leq \theta_i < t} J_i(\phi(\theta_i)), & t \geq t_0, \\ x_0 + \int_{t_0}^t f(s, \phi(s))ds - \sum_{t \leq \theta_i < t_0} J_i(\phi(\theta_i)), & t < t_0. \end{cases} \tag{2.11}$$

We shall check that $\psi \in \mathcal{PC}^1(T,\theta)$ and show that functions ψ, ϕ satisfy all conditions of Exercise 2.4.1. Condition $\phi(t_0) = \psi(t_0)$ is obviously true. If $t \notin \theta$, and it is not an end point of T, then differentiating $\psi(t)$ we find that $\psi'(t) = f(t,\phi(t)) = \phi'(t)$. We verify conditions (2) and (3) of Exercise 2.4.1 only with $t \geq t_0$. For a fix $j \in \mathcal{A}$ one has that

$$\psi(\theta_j) = x_0 + \int_{t_0}^{\theta_j} f(s,\phi(s))ds + \sum_{t_0 \leq \theta_i < \theta_j} J_i(\phi(\theta_i))$$

and

$$\psi(\theta_j - h) = x_0 + \int_{t_0}^{\theta_j - h} f(s,\phi(s))d\,s + \sum_{t_0 \leq \theta_i < \theta_j - h} J_i(\phi(\theta_i)),$$

where $h > 0$. Next, we obtain that

$$\begin{aligned} &\psi'_-(\theta_j) = \lim_{h \to 0+} \frac{\psi(\theta_j - h) - \psi(\theta_j)}{h} = \\ &\lim_{h \to 0+} \frac{1}{h}\Big[\int_{t_0}^{\theta_j - h} f(s,\phi(s))ds - \int_{t_0}^{\theta_j} f(s,\phi(s))ds\Big] = \\ &f(\theta_j, \phi(\theta_j)) = \phi'_-(\theta_j). \end{aligned} \tag{2.12}$$

If α is the right end point of T and $\alpha \in T$, then similarly one can check that $\psi'_-(\alpha) = \phi'_-(\alpha)$. Consider condition 3). Fix $j \in \mathcal{A}$, then

$$\begin{aligned} &\Delta\psi|_{t=\theta_j} = \psi(\theta_j+) - \psi(\theta_j) = \\ &\Big[x_0 + \int_{t_0}^{\theta_j+} f(s,\phi(s))ds + \sum_{t_0 \leq \theta_i < \theta_j+} J_i(\phi(\theta_i))\Big] - \\ &\Big[x_0 + \int_{t_0}^{\theta_j} f(s,\phi(s))ds + \sum_{t_0 \leq \theta_i < \theta_j} J_i(\phi(\theta_i))\Big] = \\ &J_j(\phi(\theta_j)) = \Delta\phi|_{t=\theta_j}. \end{aligned} \tag{2.13}$$

Thus, the conditions are verified if $t \geq t_0$.

Exercise 2.4.2. Verify that conditions (2), (3) of Exercise 2.4.1 are valid if $t < t_0$.

Thus, all the conditions of Exercise 2.4.1 are fulfilled and $\psi(t) = \phi(t)$. The necessity is proved. □

Sufficiency. Assume that $\phi(t)$ is a solution of (2.10). Then differentiate the expression and evaluate the discontinuities to check that it is as well a solution of (2.3). The theorem is proved. □

2.5 The Gronwall–Bellman Lemma for Piecewise Continuous Functions

Lemma 2.5.1. *Let* $u, v \in \mathcal{PC}(I, \theta), u(t) \geq 0, v(t) > 0, t \in I, \beta_i \geq 0, i \in \mathcal{A}, t_0 \in I$*, and* $c \in \mathbb{R}$ *is a nonnegative constant. From the inequality*

$$u(t) \leq c + \int_{t_0}^{t} v(s)u(s)ds + \sum_{t_0 \leq \theta_i < t} \beta_i u(\theta_i), t \geq t_0, \tag{2.14}$$

it follows that:

$$u(t) \leq c e^{\int_{t_0}^{t} v(s)ds} \prod_{t_0 \leq \theta_i < t} (1 + \beta_i), t \geq t_0. \tag{2.15}$$

Moreover, if additionally $\beta_i < 1, i \in \mathcal{A}$*, then the inequality*

$$u(t) \leq c + \int_{t_0}^{t} v(s)u(s)ds + \sum_{t \leq \theta_i < t_0} \beta_i u(\theta_i), t < t_0, \tag{2.16}$$

implies that

$$u(t) \leq c e^{-\int_{t_0}^{t} v(s)ds} \prod_{t \leq \theta_i < t_0} (1 - \beta_i)^{-1}, t < t_0. \tag{2.17}$$

Proof. The proof falls naturally in two parts. We first examine that (2.14) implies (2.15). Next, we will prove that (2.17) follows (2.16).

(a) Let $\theta_{j-1} < t_0 \leq \theta_j$, where $j \in \mathcal{A}$ is fixed. Introduce the following notations:

$$V(t) \equiv c + \int_{t_0}^{t} v(s)u(s)ds + \sum_{t_0 \leq \theta_i < t} \beta_i u(\theta_i);$$

$$W(t) \equiv c e^{\int_{t_0}^{t} v(s)ds} \prod_{t_0 \leq \theta_i < t} (1 + \beta_i).$$

To prove the assertion it is sufficient to check that

$$V(t) \leq W(t), t_0 \leq t. \tag{2.18}$$

On the interval $[t_0, \theta_j]$ inequality (2.18) is correct (see the Proof of the Gronwall–Bellman Lemma in [77]). We assume that (2.18) is true if $t \in [t_0, \theta_m]$

for some $m \in \mathcal{A}, m \geq j$, and will prove that it is fulfilled for $t \in (\theta_m, \theta_{m+1}]$. We have, for these t, the following inequalities:

$$V(t) = V(\theta_m) + \int_{\theta_m}^{t} v(s)u(s)ds + \beta_m u(\theta_m) \leq$$

$$(1 + \beta_m)V(\theta_m) + \int_{\theta_m}^{t} v(s)V(s)u(s)ds \leq$$

$$(1 + \beta_m)W(\theta_m) + \int_{\theta_m}^{t} v(s)V(s)u(s)ds.$$

Applying the Gronwall–Bellman Lemma [77] to the last inequality, one can obtain that

$$V(t) \leq (1 + \beta_m)W(\theta_m)\mathrm{e}^{\int_{\theta_m}^{t} v(s)ds} = W(t).$$

Thus, (2.15) is true by the induction.

(b) Let $\theta_j < t_0 \leq \theta_{j+1}$, where $j \in \mathcal{A}$ is fixed. Assuming $t \leq t_0$, denote

$$\tilde{V}(t) \equiv c + \int_{t_0}^{t} v(s)u(s)ds + \sum_{t \leq \theta_i < t_0} \beta_i u(\theta_i),$$

$$\tilde{W}(t) \equiv c\mathrm{e}^{-\int_{t_0}^{t} v(s)ds} \prod_{t \leq \theta_i < t_0} (1 - \beta_i)^{-1}.$$

Let us show that

$$\tilde{V}(t) \leq \tilde{W}(t), t \leq t_0, \tag{2.19}$$

to prove the assertion.

If $\theta_j < t \leq t_0$, then $u(t) \leq V(t)$, and assuming $c > 0$ (this restriction could be removed by the limit procedure), one can obtain that

$$-\frac{uv}{\tilde{V}} \geq -v.$$

Integrate the last inequality to prove

$$\tilde{V}(t) \leq \tilde{V}(t_0)\mathrm{e}^{-\int_{t_0}^{t} v(s)ds} = \tilde{W}(t).$$

Now, assume that (2.19) is true for $t \in (\theta_m, t_0], m \in \mathcal{A}, m \leq j$. Then

$$\tilde{V}(\theta_m) \leq \tilde{V}(\theta_m+) + \beta_m u(\theta_m) \leq \tilde{V}(\theta_m+) + \beta_m \tilde{V}(\theta_m)$$

or

$$\tilde{V}(\theta_m) \leq (1 - \beta_m)^{-1}\tilde{V}(\theta_m+) \leq (1 - \beta_m)^{-1}\tilde{W}(\theta_m+) = \tilde{W}(\theta_m).$$

Consequently, if $\theta_{m-1} < t < \theta_m$, then

$$\tilde{V}(t) \le \tilde{V}(\theta_m) + \int_{t_0}^{t} v(s)u(s)ds \le \tilde{W}(\theta_m) + \int_{t}^{\theta_m} v(s)\tilde{V}(s)ds.$$

That is,

$$\tilde{V}(t) \le \tilde{W}(\theta_m)e^{\int_t^{\theta_m} v(s)ds} = \tilde{W}(t).$$

Now, induction on m proves the theorem. □

Notation. Let $T \subseteq \mathbb{R}$ be an arbitrary set and $\theta \in \Theta$ a sequence. For a symbol $i(T)$, one should understand the number of elements of the sequence θ in T. For example, if $\theta_j = j, j \in \mathbb{Z}$, then $i((0,1)) = 0, i([0,1]) = 2, i([-7.5, 15]) = 23$.

Let us formulate the following two lemmas without proving them.

Lemma 2.5.2. *[142] Let $u \in \mathcal{PC}(I,\theta), t_0 \in I$, and the following inequality be valid*

$$u(t) \le \alpha + \int_{t_0}^{t} [\beta + \gamma u(s)]ds + \sum_{t_0 \le \theta_i < t} [\beta + \gamma u(\theta_i)], \tag{2.20}$$

for all $t \ge t_0$, where $\alpha \ge 0, \beta \ge 0, \gamma > 0$, are constants. Then

$$u(t) \le (\alpha + \frac{\beta}{\gamma})(1+\gamma)^{i([t_0,t])}e^{\gamma(t-t_0)} - \frac{\beta}{\gamma}. \tag{2.21}$$

Lemma 2.5.3. *Let $u \in \mathcal{PC}(I,\theta), t_0 \in I$, and the following inequality be valid:*

$$u(t) \le \alpha + \int_{t_0}^{t} [\beta + \gamma u(s)]ds + \sum_{t \le \theta_i < t_0} [\beta + \gamma u(\theta_i)], t \le t_0, \tag{2.22}$$

where $\alpha \ge 0, \beta \ge 0, 0 < \gamma < 1$, are constants. Then

$$u(t) \le (\alpha + \frac{\beta}{\gamma})(1-\gamma)^{-i([t,t_0])}e^{-\gamma(t-t_0)} - \frac{\beta}{\gamma}. \tag{2.23}$$

Exercise 2.5.1. Prove Lemmas 2.5.2 and 2.5.3.

2.6 Existence and Uniqueness Theorems

Let us consider again the impulsive differential equation (2.3), which is defined on Ω, and described in Sect. 2.1, where we clarified the concept of a solution of impulsive differential equations, and the way of extension of solutions for increasing t as well as for decreasing t. We want to determine conditions, which provide confidence that a solution of the initial value problem exists and is unique on a certain interval. Before solving this problem one should verify that the set of functions, where we are looking for a solution is complete. That is, each Cauchy sequence of functions from this set converges to a function of the same set.

Completeness of $\mathcal{PC}_r(\mathbf{T}, \boldsymbol{\theta})$. Assume that T is a bounded interval in $\mathbb{R}$. Introduce the following norm. We set $||\phi||_0 = \sup_T ||\phi(t)||$, if a function ϕ is bounded on T. Consider a Cauchy sequence of functions $\{\phi_n\} \subset \mathcal{PC}_r(T, \theta), n \geq 1$. It is easily seen that the sequence converges point-wise to a function $\phi^0(t)$ on T. Now, it is sufficient to prove that the following assertion is valid.

Theorem 2.6.1. *If* $\{\phi_n\} \subset \mathcal{PC}_r(T, \theta), n \geq 1,$ *and* $\lim_{n\to\infty} ||\phi_n - \phi^0||_0 = 0,$ *then* $\phi^0 \in \mathcal{PC}_r(J, \theta).$

Proof. Fix $i \in \mathcal{A}$ such that $(\theta_i, \theta_{i+1}] \subset T$. The functions ϕ_n are continuous on $(\theta_i, \theta_{i+1}]$ and have the right limits at $t = \theta_i$. Hence, their continuous extensions are uniformly convergent to the continuation of ϕ_0 on $[\theta_i, \theta_{i+1}]$. Consequently, the function ϕ_0 is continuous on $(\theta_i, \theta_{i+1}]$ and has the right limit at $t = \theta_i$. Similarly we can consider every interval of continuity in T. Thus, $\phi_0 \in \mathcal{PC}_r(T, \theta)$. The theorem is proved. □

Fix $(t_0, x_0) \in I \times G$, and let

$$I_0 = [t_0 - h, t_0 + h], G_0 = \{x \in \mathbb{R}^n : ||x - x_0|| < H\},$$

with some fixed positive numbers H and h.

Assume that the numbers are small such that $I_0 \times G_0 \subset I \times G$. Let $p_+ = i([t_0, t_0 + h]), p_- = i([t_0 - h, t_0]), A_0 = \{i \in \mathcal{A} : \theta_i \in I_0\}$ and $\theta^0 = \{\theta_i\}, i \in \mathcal{A}_0$.

We will make the following assumptions:

(C1) $||f(t, x) - f(t, y)|| \leq l||x - y||, ||J_i(x) - J_i(y)|| \leq l_1||x - y||$, for arbitrary $x, y \in G$, uniformly in all $(t, i) \in I \times \mathcal{A}$;

(C2) $\sup_{J\times G} ||f(t, x)|| + \sup_{\mathcal{A}\times G} ||J_i(x)|| = M < \infty$.

Theorem 2.6.2. *Let conditions (C1), (C2), and the inequalities*

$$M(h + \max(p_+, p_-)) < H, \tag{2.24}$$

$$lh + l_1 \max(p_+, p_-) < 1, \tag{2.25}$$

be valid. Then the initial value problem (2.3) and $x(t_0) = x_0$ *admit a unique solution on* I_0.

Proof. Introduce the following norm $||\phi||_1 = \sup_{I_0} ||\phi||$ for functions defined on I_0. Let

$$S_H = \{\phi \in \mathcal{PC}_r(I_0, \theta^0) : ||\phi||_1 < H\}.$$

Consider the operator $\mathcal{P}$ on S_H such that if $\phi \in S_H$, then

$$\mathcal{P}\phi(t) = \begin{cases} x_0 + \int_{t_0}^t f(s, \phi(s))ds + \sum_{t_0 \le \theta_i < t} J_i(\phi(\theta_i)), & t \ge t_0, \\ x_0 + \int_{t_0}^t f(s, \phi(s))ds - \sum_{t \le \theta_i < t_0} J_i(\phi(\theta_i)), & t < t_0. \end{cases} \tag{2.26}$$

Exercise 2.6.1. Prove that $\mathcal{P} : S_H \to S_H$.

If two functions, ϕ and ψ are from S_H then

$$\begin{aligned} ||\mathcal{P}\phi - \mathcal{P}\psi|| &\le l||\phi - \psi||_1 + l_1 \max(p_+, p_-)||\phi - \psi||_1 \\ &\le (lh + l_1 \max(p_+, p_-))||\phi - \psi||_1. \end{aligned}$$

The last inequality in view of (2.25) proves that $\mathcal{P}$ is a contraction. Consequently, there exists a unique fixed point of $\mathcal{P}$, which is a unique solution of the initial value problem. The theorem is proved. □

Exercise 2.6.2. Solve the following problems.

1. Why the completeness is needed to prove the last theorem?
2. Prove that each of the equations (2.6), where $x(t)$ is the solution determined in the last theorem, has at most one solution with respect to v.
3. Using the Schauder fixed point theorem prove that the following assertion is valid.

Theorem 2.6.3. *Assume that all conditions of the last theorem are true, except inequality (2.25). Then (2.3) admits a solution defined on I_0.*

2.7 Continuity

In this section we assume that $I \times G$ is a bounded set in $\mathbb{R} \times \mathbb{R}^n$, and fix a finite and closed interval $\tilde{I} = [a, b] \subset I$. We consider a solution $x(t) = x(t, t_0, x_0), t_0, t \in \tilde{I}$, of (2.3) with moments of discontinuity $\theta_i, a < \theta_m < \ldots < \theta_k < b$. That is, we suppose that the moments of discontinuity are interior points of the section. Denote by $G_{m-1} = \{(t, x) : a \le t \le \theta_m, x \in G\}, G_k = \{(t, x) : \theta_k < t \le b, x \in G\}$ and $G_i = \{(t, x) : \theta_i < t \le \theta_{i+1}, x \in G\}, i = m, \ldots, k-1$.

Definition 2.7.1. The solution $x(t) = x(t, t_0, x_0)$ of (2.3) continuously depends on x_0 if to any $\epsilon > 0$ there corresponds $\delta > 0$ such that any other solution $\tilde{x}(t)$ with $||x_0 - \tilde{x}(t_0)|| < \delta$ is continuable on $\tilde{I}$, and $||x(t) - \tilde{x}(t)|| < \epsilon$ on this interval.

Denote by $d(x_0, \delta) = \{(t, x) \in \mathbb{R} \times \mathbb{R}^n : t_0 - \delta < t < t_0 + \delta, x = x_0\}$ the set, where δ is a positive real number, and fix j such that $(t_0, x_0) \in G_j$.

Definition 2.7.2. The solution $x(t) = x(t, t_0, x_0)$ of (2.3) continuously depends on t_0 if to any $\epsilon > 0$ there corresponds $\delta > 0$ such that any other solution $\tilde{x}(t) = x(t, \tilde{t}_0, x_0)$ is continuable on $\tilde{I}$, and $||x(t) - \tilde{x}(t)|| < \epsilon$ on this interval, as soon as $\tilde{t}_0 \in d(x_0, \delta) \cap G_j$.

Consider the following system:

$$\begin{aligned} y' &= f(t,y) + g(t,y), \\ \Delta y|_{t \neq \theta_i} &= J_i(y) + W_i(y), \end{aligned} \tag{2.27}$$

which is defined as well as (2.3) on Ω, and assume that function g is continuous, functions $W_i, i \in \mathcal{A}$, are defined on G, and

(C3) $\sup_{I\times G} ||g|| + \sup_{\mathcal{A}\times G} ||W|| = \eta < \infty.$

Definition 2.7.3. The solution $x(t) = x(t, t_0, x_0)$ of (2.3) continuously depends on the right-hand side of the system if to any $\epsilon > 0$ there corresponds $\delta > 0$ such that any solution $y(t)$ of (2.27) with $y(t_0) = x_0$ is continuable on $\tilde{I}$, and $||x(t) - y(t)|| < \epsilon$ on this interval, whenever $\eta < \delta$.

Lemma 2.7.1. *Suppose functions f, J_i, satisfy condition (C1) with $l_1 = l$, functions g, W are such that $(C3)$ is valid, and a solution $y(t) = y(t, t_0, y_0)$ of (2.27) exists on $\tilde{I}$.*

Then the inequality

$$||x(t) - y(t)|| \le (||x_0 - y_0|| + \frac{\eta}{l})(1+l)^{i([t_0,t])}e^{l(t-t_0)} - \frac{\eta}{l}, \tag{2.28}$$

is true, where $t \geq t_0$. Moreover, if $l < 1$, then

$$||x(t) - y(t)|| \le (||x_0 - y_0|| + \frac{\eta}{l})(1-l)^{-i([t,t_0])}e^{-l(t-t_0)} - \frac{\eta}{l}, \tag{2.29}$$

where $t \leq t_0$.

Proof. Consider (2.29). The case with $t \geq t_0$ could be discussed similarly. Let $y(t) = y(t, t_0, y_0)$ be a solution of (2.27) defined on $\tilde{I}$. Then for $t \leq t_0$ we have that

$$y(t) = y_0 + \int_{t_0}^{t} \Big[f(s,y(s)) + g(s,y(s))\Big]ds - \sum_{t\le\theta_i<t_0} J_i(y(\theta_i)). \tag{2.30}$$

Moreover,

$$x(t) = x_0 + \int_{t_0}^{t} f(s,x(s))ds - \sum_{t\le\theta_i<t_0} J_i(x(\theta_i)). \tag{2.31}$$

Finding difference of the last two equalities one can obtain, through conditions (C1), (C3), that

$$\begin{aligned} ||x(t) - y(t)|| \le ||x_0 - y_0|| + \int_{t_0}^{t} [\eta + l||x(s) - y(s)||]ds \\ + \sum_{t\le\theta_i<t_0} [\eta + l||x(\theta_i) - y(\theta_i)||]. \end{aligned}$$

Applying Lemma 2.5.1 to the last inequality, we find that (2.29) is true. The lemma is proved. □

Lemma 2.7.2. *Suppose* $||J_i(x) - J_i(y)|| \leq l_1||x - y||$ *for all* $i \in \mathcal{A}, x, y \in G$. *If* $l_1 < 1$, *then the equation*

$$x = v + J_i(v),$$

$i \in \mathcal{A}, x \in G$, *has at most one solution* $v \in G$. *Moreover, if* $x_j = v_j + J_i(v_j), j = 1, 2,$ *then*

$$||v_1 - v_2|| \leq \frac{1}{1 - l_1}\|x_1 - x_2\|. \tag{2.32}$$

Proof. Fix $i \in \mathcal{A}$. We have that $\|x_1 - x_2\| = \|v_1 - v_2 + J_i(v_1) - J_i(v_2)\| \geq \|v_1 - v_2\| - l_1\|v_1 - v_2\| = (1 - l_1)\|v_1 - v_2\|$. Consequently, (2.32) is correct. The uniqueness follows this formula immediately. The lemma is proved. □

Theorem 2.7.1. *Suppose* f, J *satisfy (C1) and* $l_1 < 1$. *Then the solution* $x(t)$ *of (2.3) continuously depends on* x_0 *if* $t_0 = b$. *If* $\tilde{x}(t) = x(t, t_0, \tilde{x}_0)$ *is another solution of (2.3), and* $||x_0 - \tilde{x}_0||$ *is sufficiently small, then*

$$||x(t) - \tilde{x}(t)|| \leq ||x_0 - \tilde{x}_0||(1 - l_1)^{-i([t,t_0])}e^{-l(t-t_0)}, \quad a \leq t \leq t_0. \tag{2.33}$$

Proof. Fix $\epsilon > 0$. One can take δ' sufficiently small such that the tube $\mathcal{T} = \{(t, x) : \|x - x(t)\| < \delta', t \in [a, t_0]\}$ is entirely in $I \times G$. Moreover, by Lemma 2.7.2, a number $\delta' > 0$ can be chosen such that the maps Π_i^{-1} are defined in the small neighborhoods of $x(\theta_i+)$.

Fix a positive δ such that $\delta(1 - l_1)^{-i([a,t_0])}e^{l(t_0-a)} < \min\{\delta', \epsilon\}$. Let $\tilde{x}(t) = x(t, t_0, \tilde{x}_0)$ be another solution of (2.3), and $||x_0 - \tilde{x}_0|| < \delta$. Applying Lemma 2.5.1 and formula (2.29), we obtain inequality (2.33). Since of Theorem 2.4.1, Lemma 2.7.2, and condition (M0), the solution $\tilde{x}(t)$ is continuable on $\tilde{I}$ and $||x(t) - \tilde{x}(t)|| < \epsilon$ for all $t \in [a, t_0]$. The theorem is proved. □

We will formulate several theorems on the continuous dependence without proving them.

Theorem 2.7.2. *Suppose functions* f, J *satisfy (C1). Then the solution* $x(t)$ *of (2.3) continuously depends on* x_0 *if* $t_0 = a$.

If $\tilde{x}(t) = x(t, t_0, \tilde{x}_0)$ *is another solution of (2.3), and* $||x_0 - \tilde{x}_0||$ *is sufficiently small, then*

$$||x(t) - \tilde{x}(t)|| \leq ||x_0 - \tilde{x}_0||(1 + l_1)^{i([t_0,t))}e^{l(t-t_0)}, \quad t_0 \leq t \leq b. \tag{2.34}$$

Theorem 2.7.3. *Suppose* f, J *satisfy (C1), and* $l_1 < 1$. *Then every solution* $x(t) = x(t, t_0, x_0), (t_0, x_0) \in I \times G$, *of (2.3) continuously depends on* x_0 *on any closed and finite interval, where it is defined.*

By reduction to the dependence on the initial value one can prove that the following assertions are valid.

Theorem 2.7.4. *Assume that functions f, J satisfy (C1). Then the solution $x(t)$ of (2.3) continuously depends on t_0 if $t_0 = a$.*

Theorem 2.7.5. *Assume that f, J satisfy (C1) and $l_1 < 1$. Then the solution $x(t) = x(t, t_0, x_0)$ of (2.3) continuously depends on t_0 if $t_0 = b$.*

Theorem 2.7.6. *Assume that f, J satisfy (C1) and $l_1 < 1$. Then every solution $x(t) = x(t, t_0, x_0), (t_0, x_0) \in I \times G$, of (2.3) continuously depends on t_0 on each closed and finite interval, where it is defined.*

Exercise 2.7.1. Explain, why the set $d(x_0, \delta)$ in Definition 2.7.2 cannot be ignored in the last three theorems. Give a simple example, where a solution with initial discontinuity moment does not depend continuously on the initial moment despite the equation 'encourages' the continuity.

Hint: Consider the following system:

$$\begin{aligned} x' &= 0, \\ \Delta x|_{t \neq i} &= 1, \end{aligned} \tag{2.35}$$

where $x, t \in \mathbb{R}, i \in \mathbb{Z}$, and the solution $x(t, 0, 0)$ of this system if $t > 0$.

Theorem 2.7.7. *Assume that f, J satisfy (C1), $l_1 = l < 1$, functions g, W satisfy $(C2)$ and $||g(t,x) - g(t,y)|| + ||W_i(x) - W_i(y)|| \leq l_2||x - y||, l_2$ is a positive constant, for arbitrary $x, y \in G$ uniformly in all $(t, i) \in I \times \mathcal{A}$. Then the solution $x(t)$ of (2.3) continuously depends on the right-hand side of the system.*

Exercise 2.7.2. Prove Theorems 2.7.2 to 2.7.7. Hint: To prove Theorem 2.7.7, use Lemma 2.7.1.

Remark 2.7.1. In Theorem 2.7.7 the Lipschitz condition for g, W can be replaced by the local existence assumption.

Remark 2.7.2. Assume that function $f(t, x)$ in (2.3) satisfies a Lipschitz condition, and all transition operators Π_i are homeomorphisms. Then one can see that the solution $x(t, t_0, x_0)$ of (2.3) continuously depends on the initial condition. Theorems with strong restrictions on the impulsive functions are important for applications as they provide exact evaluations of the dependence.

Notes

To the best of our knowledge, the history of mathematical treatment of discontinuous processes began in the middle of the last century. Various aspects of the theory and its applications can be found in pioneer papers and books [38, 42, 75, 78, 90, 91, 107, 116, 143, 147–149, 154]. In [111], the authors expressed the idea of general theory of differential equations with fixed moments of impulses. Papers [137–140] developed this idea, and first theorems, analogues to those for ordinary differential equations, were proved. The investigation was continued in [22, 30, 45, 102, 128] and many other papers. These results were summarized in the books [95, 141]. Further, extended version of [141] was published in English [142]. Some results of the present chapter can be found in [46, 95, 111, 137–140, 142]. The Gronwall–Bellman Lemma for piecewise continuous functions is published in [138], and in [46] the version with $t \leq t_0$ is proved. The importance of the transition operator for continuation of solutions was first mentioned in [111]. The local existence and uniqueness theorems, different versions of extension theorems, including assertions on maximal intervals of existence, continuous dependence of solutions on initial conditions are considered in [46, 111, 142]. In the present book, Theorems 2.3.3, 2.6.2, 2.6.3, 2.7.1, 2.7.5, 2.7.6, 2.7.7, of global existence and uniqueness, extension of solutions, maximal intervals of existence, continuous dependence on initial conditions and the right-hand side, when the time is increasing and decreasing, are formulated in the unified form, similarly to ordinary differential equations. The form is convenient for searching new sufficient conditions, which may provide the qualitative properties required by future applications. Theorems 2.3.1, 2.3.3, 2.4.1, 2.7.1, 2.7.3, 2.7.6, 2.7.7, as well as Definition 2.7.2, which uses the set-interval $d(x_0, \delta)$, are published for the first time.

In addition, we want to make the following two important remarks:

1. It is the first time that the differential equation with impulses is given in the form

$$\begin{aligned} x' &= f(t,x), \\ \Delta x|_{t=\theta_i} &= J_i(x), \end{aligned} \tag{2.36}$$

   in the present book. Commonly the system

$$\begin{aligned} x' &= f(t,x), t \neq \theta_i, \\ \Delta x|_{t=\theta_i} &= J_i(x), \end{aligned} \tag{2.37}$$

   has been used [95, 142]. The system (2.36) is more adequate to the motion with discontinuities than (2.37), since it does not ignore the existence of the left derivative at the points of discontinuity.
   We should say to the reader that system (2.36) can be accepted as (2.37) and vise versa. So, one can read our book considering (2.36) as equation of the type (2.37), for convenience. That is, assuming that inequality $t \neq \theta_i$ is involved in the first line of (2.36). Moreover, introduction of the new form, (2.36), changes nothing in all proofs of the theory.
2. In [95] (see also [46]), the initial condition $x(t_0+) = x_0$ is considered. This excludes a jump at the initial moment, when a solution is continued to the right. In our book we keep the initial condition in its classical form $x(t_0) = x_0$, [142]. Thus, we have achieved the maximal parallelism with ordinary differential equations theory in proofs as well as in notations.

Chapter 3
Stability and Periodic Solutions of Systems with Fixed Moments of Impulses

3.1 Definitions of Stability

Let us consider a differential equation with impulses,

$$x'(t) = f(t,x),$$
$$\Delta x|_{t=\theta_i} = J_i(x), \qquad (3.1)$$

which is defined on the set Ω. In this section we assume that $I = [0,\infty)$, and $\theta_i \to \infty$ as $i \to \infty$. Let $\phi(t)$ be a solution of (3.1) such that $\phi : I \to G$.

Definition 3.1.1. The solution $\phi(t)$ is stable if to any $\epsilon > 0$ and $t_0 \in I$ there corresponds $\delta(t_0,\epsilon) > 0$ such that for any other solution $\psi(t)$ of (3.1) with $||\phi(t_0) - \psi(t_0)|| < \delta(t_0,\epsilon)$ we have $||\phi(t) - \psi(t)|| < \epsilon$ for $t \geq t_0$.

Definition 3.1.2. The solution $\phi(t)$ is uniformly stable, if $\delta(t_0,\epsilon)$ from Definition 3.1.1 can be chosen independently of t_0.

Definition 3.1.3. The solution $\phi(t)$ is asymptotically stable if it is stable in the sense of Definition 3.1.1 and there exists a positive number $\kappa(t_0)$ such that if $\psi(t)$ is any other solution of (3.1) with $||\phi(t_0) - \psi(t_0)|| < \kappa(t_0)$, then $||\phi(t) - \psi(t)|| \to 0$ as $t \to \infty$.

Definition 3.1.4. The solution $\phi(t)$ is uniformly asymptotically stable if it is uniformly stable in the sense of Definition 3.1.2, and given $\epsilon > 0$ we can find $T(\epsilon) > 0$ such that for any other solution $\psi(t)$ of (3.1) with $||\phi(t_0) - \psi(t_0)|| < \kappa$, where κ is independent of t_0, we have $||\phi(t) - \psi(t)|| < \epsilon$ for $t \geq t_0 + T(\epsilon)$.

Definition 3.1.5. The solution $\phi(t)$ is unstable if there exist numbers $\epsilon_0 > 0$ and $t_0 \in I$ such that for any $\delta > 0$ there exists a solution $y_\delta(t)$, $||\phi(t_0) - y_\delta(t_0)|| < \delta$, of (3.1) such that either it is not continuable to ∞ or there exists a moment $t_1, t_1 > t_0$ such that $||y_\delta(t_1) - \phi(t_1)|| \geq \epsilon_0$.

M. Akhmet, *Principles of Discontinuous Dynamical Systems*,
DOI 10.1007/978-1-4419-6581-3_3, © Springer Science+Business Media, LLC 2010

Example 3.1.1. Consider the following system:

$$\begin{aligned} x' &= -2x, \\ \Delta x|_{t=i} &= \alpha x, \end{aligned} \tag{3.2}$$

where $i \in \mathbb{Z}$.

Define the coefficient $\alpha \in \mathbb{R}$ such that the trivial solution $x \equiv 0$ of the system is: (a) stable; (b) asymptotically stable; (c) unstable.

Solution. Fix $j \in \mathbb{Z}, j \geq 0$. One can find that

$$x(j+1) = x(j)\mathrm{e}^{-2}(1+\alpha).$$

Set $q = |\mathrm{e}^{-2}(1+\alpha)|$. Applying the last formula, we can obtain by recursion that

$$|x(i)| = |x_0|q^i, i \geq 1.$$

The last formula implies easily that:

(i) if $q = 1$ then $|x(i)| = |x_0|, i \geq 1$;
(ii) if $q < 1$ then $|x(i)| \to 0$ as $i \to \infty$;
(iii) if $q > 1$ then $|x(i)| \to \infty$ as $i \to \infty$.

Now, by using the cases (i)–(iii), we shall prove that the zero solution is stable if $\alpha = -1 \pm e^2$, asymptotically stable if $-1 - e^2 < \alpha < -1 + e^2$, and unstable if $\alpha < -1 - e^2$ or $\alpha > -1 + e^2$.

Indeed, if $\alpha = -1 \pm e^2$, then $q = 1$. Fix $\epsilon > 0$ and choose a positive δ such that $\delta < \frac{\epsilon}{|1+\alpha|}$.

Let us show that $|x_0| < \delta$ implies $|x(t, 0, x_0)| < \epsilon, t \geq 0$. Since $|\mathrm{e}^{-2}(1+\alpha)| = 1$, we have that $|1+\alpha| > 1$ and, consequently,

$$|x(i)| = |x_0| < \delta < \frac{\epsilon}{|1+\alpha|} < \epsilon, i \geq 0.$$

Moreover, the equality $x(i+) = x(i)(1+\alpha)$ implies that $|x(i+)| = |x(i)||1+\alpha| < \delta|1+\alpha| < \epsilon$.

On each interval $(i, i+1], i \geq 0$, the expression $|x(t)|$ is a decreasing function. Hence, $|x(t)| \leq |x(i+)| < \epsilon$ if $i < t \leq i+1$. The stability of the trivial solution is proved.

Exercise 3.1.1. Prove that the trivial solution is stable if $q < 1$.

Assume, now, that $-1 - e^2 < \alpha < -1 + e^2$. Then $q < 1$. We have that $|x(i+)| = |x_0|q^i(1+\alpha)$, $i \geq 0$, and, hence, $x(i+) \to 0$ as $i \to \infty$. Moreover, $|x(t)|$ is a decreasing function on each interval $(i, i+1], i \geq 0$. That is, $x(t) \to 0$ as $t \to \infty$. Since the last relation is valid for an arbitrary x_0, the zero solution is asymptotically stable in large.

If $\alpha < -1 - e^2$ or $\alpha > -1 + e^2$, then $q > 1$. Fix $\epsilon > 0$ and an arbitrary $\delta, \delta < \epsilon$. Consider the solution $x(t) = x(t, 0, x_0)$ such that $|x_0| = \frac{\delta}{2}$. If $i_0 = \left[\frac{\ln \frac{2\epsilon}{\delta}}{\ln q}\right] + 1$, where $[\cdot]$ is the greatest integer function, then $|x(i)| = \frac{\delta}{2} q^i$ implies that $|x(i)| \geq \epsilon$, $i \geq i_0$. Thus, the trivial solution is unstable.

3.2 Basics of Periodic Systems

Consider system (3.1) assuming that $I = \mathbb{R}, \mathcal{A} = \mathbb{Z}$, and $|\theta_i| \to \infty$ as $|i| \to \infty$. Fix positive $\omega \in \mathbb{R}, p \in \mathbb{Z}$.

We shall need the following additional conditions of periodicity:

(C4) $f(t + \omega, x) = f(t, x)$ for all $(t, x) \in \mathbb{R} \times G$;
(C5) $J_{i+p}(x) = J_i(x)$ for all $(i, x) \in \mathbb{Z} \times G$;
(C6) $\theta_{i+p} = \theta_i + \omega$ for all $i \in \mathbb{Z}$, (p-property).

If conditions (C4)–(C6) are valid, then (3.1) is called an *(ω, p)-periodic system.*

Exercise 3.2.1. Prove that the system

$$\begin{aligned} x' &= -2\cos(\pi t)x^2, \\ \Delta x|_{t=\theta_i} &= e^{2(-1)^i x}, \end{aligned} \tag{3.3}$$

where $\theta_i = i + (-1)^{i+1}\frac{1}{3}, i \in \mathbb{Z}$, is $(2, 2)$-periodic.

A function $\phi(t) \in \mathcal{PC}(\mathbb{R}, \theta)$ is a piecewise continuous ω-periodic function (or just a periodic function) if

$$\phi(t + \omega) = \phi(t),$$

for all $t \in \mathbb{R}$. Denote by $\mathcal{PC}_p$, the set of all periodic functions from $\mathcal{PC}(\mathbb{R})$.

Exercise 3.2.2. If $\phi(t) \in \mathcal{PC}_p$ and the set of discontinuity moments θ of this function is not empty, then prove that there exist a positive real number ω and an integer p such that $(C6)$ is valid.

Exercise 3.2.3. Prove that every function $\phi(t) \in \mathcal{PC}_p$:

(a) is bounded on $\mathbb{R}$;
(b) is uniformly continuous on the union of intervals of continuity;
(c) satisfies $\phi(t + k\omega) = \phi(t)$, for all $(t, k) \in \mathbb{R} \times \mathbb{Z}$, if ω is its period.

Lemma 3.2.1. *If $\phi(t) \in \mathcal{PC}(\mathbb{R}, \theta)$ is a solution of (ω, p)-periodic system (3.1), then $\phi(t + k\omega), k \in \mathbb{Z}$, also satisfies the system.*

Proof. Denote $\psi(t) = \phi(t + k\omega), k \in \mathbb{Z}$. We have that

$$\frac{d\psi(t)}{dt} = \frac{d\phi(t + k\omega)}{d(t + k\omega)} \frac{d(t + k\omega)}{dt} = f(t + k\omega, \phi(t + k\omega)) = f(t, \psi(t)).$$

One can also see that the jump equation is satisfied. Indeed,

$$\Delta\psi|_{t=\theta_i} = \psi(\theta_i+) - \psi(\theta_i) = \phi(\theta_i + k\omega+) - \phi(\theta_i + k\omega) =$$

$$\phi(\theta_{i+kp}+) - \phi(\theta_{i+kp}) = J_{i+kp}(\phi(\theta_{i+kp})) = J_i(\phi(\theta_i + k\omega)) = J_i(\psi(\theta_i)).$$

The lemma is proved. □

Theorem 3.2.1. *(Poincaré criterion) Assume that conditions (C4)–(C6) are valid and $f(t,x)$ satisfies a local Lipschitz condition. A solution $\phi(t) \in \mathcal{PC}(\mathbb{R},\theta)$ of (3.1) is ω-periodic if and only if $\phi(0) = \phi(\omega)$.*

Proof. Necessity of the theorem is obvious. Let us prove sufficiency. Write $\psi(t) = \phi(t+\omega)$. By Lemma 3.2.1, $\psi(t)$ is a solution of (3.1). Moreover, $\psi(0) = \phi(0) = \phi(\omega)$. That is, ψ and ϕ are solutions with one and the same initial values. Consequently, $\psi(t) = \phi(t), t \in \mathbb{R}$, i.e., $\phi(t+\omega) = \phi(t), t \in \mathbb{R}$. The theorem is proved. □

Exercise 3.2.4. Explain, why we do not impose any conditions on the impulsive functions J in the last theorem.

Example 3.2.1. For system (3.2), find values of the coefficient α such that all solutions are 1-periodic. Investigate stability of these solutions.

Solution. Using the formula $x(i) = x_0 q^i, i \geq 0$, and the Poincaré criterion one can obtain that all solutions of (3.2) are 1-periodic functions if and only if

$$x_0 = x_0 \mathrm{e}^{-2}(1+\alpha).$$

That is, all solutions of the system

$$\begin{aligned} x' &= -2x, \\ \Delta x|_{t=i} &= (\mathrm{e}^2 - 1)x \end{aligned} \tag{3.4}$$

are 1-periodic.

Let $\phi(t)$ be a 1-periodic solution of (3.4) and $\psi(t)$ be another solution of the system. Verify that the difference $\phi(t) - \psi(t)$ is also a solution of (3.4). Since the trivial solution of (3.4) is stable, solution $\phi(t)$ is also stable.

Exercise 3.2.5. By using the periodicity of all solutions and the continuous dependence of solutions on the initial value prove that the solution $\phi(t)$ is stable.

Example 3.2.2. Investigate periodicity and stability of solutions of the following system:

$$\begin{aligned} x' &= 0, \\ \Delta x|_{t=i} &= -x + \frac{1}{x}, \end{aligned} \tag{3.5}$$

where $(t,x) \in \mathbb{R} \times \mathbb{R}$.

Solution. It is not difficult to see that

$$x(i+1) = \frac{1}{x(i)}, \tag{3.6}$$

if $x(t)$ is a solution such that $x(i) \neq 0, i \in \mathbb{Z}$.

Let $x(t) = x(t, t_0, 0)$ be a solution of (3.6) with $j-1 < t_0 \leq j$ for some $j \in \mathbb{Z}$. We have that $x(t) = 0$, if $t \in [t_0, j]$ and it is not continuable beyond $t = j$.

Consider solutions with a nonzero initial value. Fix $x_0 > 0$. One can verify that

$$x(t) = \begin{cases} x_0, & t_0 \leq t \leq j, \\ \frac{1}{x_0}, & j < t \leq j+1, \\ x_0, & j+1 < t \leq j+2. \end{cases} \tag{3.7}$$

is a solution such that $x(t_0) = x_0$. Proceeding, one can see that $x(t)$ is continuable as a 2-periodic solution, and

$$x(t) = \begin{cases} \frac{1}{x_0}, & j+2i < t \leq j+2i+1, \\ x_0, & j+2i+1 < t \leq j+2(i+1), \end{cases} \tag{3.8}$$

where $i \in \mathbb{Z}$.

Thus, every solution with a positive initial value is a 2-periodic function. Let us investigate stability of this solution. Assume that $\bar{x}(t) = x(t, 0, \bar{x}_0)$ is a neighbor solution. Fix $\epsilon > 0$, and choose

$$\delta = \min\{\epsilon, \frac{x_0}{2}, \frac{\epsilon x_0^2}{1+\epsilon x_0}\}.$$

Let $|x_0 - \bar{x}_0| < \delta$. Then $\bar{x}_0 > x_0 - \delta > 0$.

1. If $t_0 \leq t \leq j$, then

$$|x(t) - \bar{x}(t)| = |x_0 - \bar{x}_0| < \delta \leq \epsilon.$$

2. If $j < t \leq j+1$, then

$$|x(t) - \bar{x}(t)| = |\frac{1}{x_0} - \frac{1}{\bar{x}_0}| = \frac{|x_0 - \bar{x}_0|}{x_0 \bar{x}_0} < \frac{\delta}{x_0(x_0 - \delta)} \leq \epsilon.$$

3. If $j+1 < t \leq t_0 + 2$, then again we have that

$$|x(t) - \bar{x}(t)| = |x_0 - \bar{x}_0| < \delta \leq \epsilon.$$

Now, the periodicity of $x(t)$ and $\bar{x}(t)$ implies that $|x(t) - \bar{x}(t)| < \epsilon$ for all $t \geq t_0$. That is, $x(t)$ is a stable solution.

Exercise 3.2.6. Verify that a solution $x(t) = x(t, t_0, x_0), x_0 < 0$, of (3.5) is continuable on $\mathbb{R}$, 2-periodic and stable.

Exercise 3.2.7. [141] Consider the following system:

$$\begin{aligned} x' &= 1 + x^2, \\ \Delta x|_{t=\theta_i} &= -1, \end{aligned} \tag{3.9}$$

where $\theta_i = i\frac{\pi}{4}, i \in \mathbb{Z}$. Prove that:

(i) the solution $x(t, 0, 1)$ of (3.9) is $\frac{\pi}{4}$-periodic;
(ii) every solution of (3.9) is unstable.

Notes

The main results of the chapter were published in [137–141].

Chapter 4
Basics of Linear Systems

4.1 Linear Homogeneous Systems

We start discussion of linear impulsive systems with the following differential equation:

$$\begin{aligned} x' &= A(t)x, \\ \Delta x|_{t=\theta_i} &= B_i x, \end{aligned} \tag{4.1}$$

where $(t, x) \in \mathbb{R} \times \mathbb{R}^n$, $\theta_i, i \in \mathbb{Z}$, is a B-sequence, such that $|\theta_i| \to \infty$ as $|i| \to \infty$. We suppose that the entries of $n \times n$ matrix $A(t)$ are from $\mathcal{PC}(\mathbb{R}, \theta)$, real valued $n \times n$ matrices $B_i, i \in \mathbb{Z}$, satisfy

$$\det(\mathcal{I} + B_i) \neq 0, \tag{4.2}$$

where $\mathcal{I}$ is the identical $n \times n$ matrix.

Theorem 4.1.1. *Every solution $x(t) = x(t, t_0, x_0), (t_0, x_0) \in \mathbb{R} \times \mathbb{R}^n$, of (4.1) is unique and continuable on $\mathbb{R}$.*

Proof. There exists $j \in \mathbb{Z}$ such that $\theta_{j-1} < t_0 \leq \theta_j$. Assume that $t_0 < \theta_j$. Since $A(t) \in \mathcal{PC}(\mathbb{R}, \theta)$, the matrix is continuous on $[t_0, \theta_j]$. That is why, the solution $x(t)$ exists and unique on this interval as the solution $\phi(t, t_0, x_0)$ of the linear homogeneous ordinary differential equation

$$x' = A(t)x. \tag{4.3}$$

If $t_0 = \theta_j$, then the solution starts with a jump. Next, we have that $x(\theta_j+) = (\mathcal{I} + B_j)x(\theta_j)$. Construct the following system (see Remark 2.3.1):

$$x' = A_e(t)x, \tag{4.4}$$

where $A_e(t) = A(t)$ if $t \neq \theta_i$, and $A_e(\theta_i) = A(\theta_i+), i \in \mathbb{Z}$.

M. Akhmet, *Principles of Discontinuous Dynamical Systems*,
DOI 10.1007/978-1-4419-6581-3_4, © Springer Science+Business Media, LLC 2010

Solution $x(t)$ is equal to the solution $\psi(t, \theta_j, x(\theta_j+))$ of (4.4) on the interval $(\theta_j, \theta_{j+1}]$. The solution exists, unique and continuable on $(\theta_j, \theta_{j+1}]$. Similarly we can obtain the proof for all $t \geq t_0$.

Let us consider $t \leq t_0$. Equation (4.3) has the unique solution $\phi(t, t_0, x_0)$ on the interval $(\theta_{j-1}, t_0]$, and $x(t) = \phi(t, t_0, x_0)$. Obviously, we have that $x(\theta_{j-1}+) = \phi(\theta_{j-1}, t_0, x_0)$. Solving the equation

$$x(\theta_{j-1}+) = (\mathcal{I} + B_{j-1})x(\theta_{j-1}),$$

one can find

$$x(\theta_{j-1}) = (\mathcal{I} + B_{j-1})^{-1}x(\theta_{j-1}+).$$

Next, we have that $x(t) = \phi(t, \theta_{j-1}, x(\theta_{j-1}))$ on the interval $(\theta_{j-2}, \theta_{j-1}]$, where $\phi(t, \theta_{j-1}, x(\theta_{j-1}))$ is the unique solution of (4.3). Proceeding in this way one can complete the proof for all $t \leq t_0$. The theorem is proved. □

Exercise 4.1.1. Consider system (4.1) with $A(t) \in \mathcal{PC}_r(J, \theta)$, $J \subset \mathbb{R}$. Prove that every solution $x(t) = x(t, t_0, x_0)$, $(t_0, x_0) \in J \times \mathbb{R}^n$, of (4.1) is unique, continuable on J, and $x(t) \in \mathcal{PC}_r^1(J, \theta)$.

Hint: Use Theorem 4.1.1, Definitions 2.1.4, 2.1.5, and Exercise 2.3.1.

Exercise 4.1.2. Suppose that the entries of $A(t)$ belong to $\mathcal{PC}(J, \theta)$, $J \subset \mathbb{R}$. Prove that every solution $x(t) = x(t, t_0, x_0)$, $(t_0, x_0) \in J \times \mathbb{R}^n$, of (4.1) is unique, continuable on J, and $x(t) \in \mathcal{PC}^1(J, \theta)$.

Lemma 4.1.1. *The set $\mathcal{X}$ of all solutions of (4.1) is a linear space.*

Proof. Let $x_1(t), x_2(t) : \mathbb{R} \to \mathbb{R}^n$, be two solutions of (4.1). We shall show that $\alpha x_1(t) + x_2(t), \alpha \in \mathbb{R}$, is also a solution of (4.1).

Indeed, we first have that

$$\begin{aligned}(\alpha x_1(t) + x_2(t))' &= \alpha x_1'(t) + x_2'(t) = \alpha A(t)x_1(t) + A(t)x_2(t)\\ &= A(t)(\alpha x_1(t) + x_2(t)).\end{aligned}$$

It means that the linear combination satisfies the differential equation. Moreover, for a fixed $i \in \mathbb{Z}$ one can obtain that

$$\Delta(\alpha x_1 + x_2)|_{t=\theta_i} = \alpha x_1(\theta_i+) + x_2(\theta_i+) - \alpha x_1(\theta_i) - x_2(\theta_i) =$$

$$\alpha\Delta x_1|_{t=\theta_i} + \Delta x_2|_{t=\theta_i} = \alpha B_i x_1(\theta_i) + B_i x_2(\theta_i) = B_i(\alpha x_1(\theta_i) + x_2(\theta_i)).$$

Thus, the equation of jumps is also satisfied by the linear combination of solutions. The lemma is proved. □

Lemma 4.1.2. *The linear space $\mathcal{X}$ has dimension n.*

The proof of this lemma is the same as that of Theorem 4.2 [60].

Let $x_i(t), i = 1, 2, \ldots, n$, be a basis of the space $\mathcal{X}$. It is called a *fundamental system of solutions* of (4.1).

Set $x_i(t) := (x_i^1, x_i^2, \ldots, x_i^n), i = 1, 2, \ldots, n$, and consider the matrix

$$X(t) = \begin{pmatrix} x_1^1 & x_2^1 \ldots x_n^1 \\ x_1^2 & x_2^2 \ldots x_n^2 \\ \ldots & \\ x_1^n & x_2^n \ldots x_n^n \end{pmatrix}.$$

The matrix $X(t)$ is called a *fundamental matrix of solutions* (a fundamental matrix) of (4.1).

Exercise 4.1.3. Prove the following theorem.

Theorem 4.1.2. *A set of solutions $x_1(t), x_2(t), \ldots, x_n(t)$, of (4.1) is a fundamental system if and only if* $\det X(t) \neq 0$, *for all $t \in \mathbb{R}$.*

Denote by $X(t,s), t, s \in \mathbb{R}$, a fundamental matrix of (4.1) such that $X(s,s) = \mathcal{I}$, $s \in \mathbb{R}$. We shall call it a *transition matrix.*

Exercise 4.1.4. Let $X(t)$ be a fundamental matrix of (4.1). Show that $X(t,s) = X(t)X^{-1}(s)$ for all $t, s \in \mathbb{R}$.

It is useful to to solve the following two problems.

Example 4.1.1. Construct a fundamental matrix of (4.1) using the transition matrix $U(t,s)$ of (4.3) and matrices $B_i, i \in \mathbb{Z}$.

Solution. Since all solutions of (4.1) are defined on $\mathbb{R}$ and the solution of the initial value problem is unique, assume, without loss of generality, that $t_0 = 0$ and fix $k \in \mathbb{Z}$ such that $\theta_{k-1} < 0 \le \theta_k$. Denote $X_0 = X(0)$.

Using the extension procedure of solutions, which is described in Sect. 2.3, one can find that

$$X(t) = \begin{cases} X_0, if\ t = 0, \\ U(t,\theta_p)(\mathcal{I} + B_p)U(\theta_p, \theta_{p-1}) \ldots \\ U(\theta_{k+1}, \theta_k)(\mathcal{I} + B_k)U(\theta_{k-1}, t_0)X_0, if\ \theta_k \le \theta_p < t \le \theta_{p+1}, \\ U(t,t_0)X_0, if\ t_0 \le t \le \theta_k, \\ U(t,\theta_l)(\mathcal{I} + B_l)^{-1}U(\theta_l, \theta_{l+1}) \ldots \\ U(\theta_{k-1}, \theta_{k-1})(\mathcal{I} + B_{k-1})^{-1}U(\theta_{k-1}, 0)X_0, if\ \theta_{l-1} < t \le \theta_l \le \theta_{k-1}, \\ U(t,t_0)X_0, if\ \theta_{k-1} < t \le 0. \end{cases} \tag{4.5}$$

From the last formula it is not difficult to see that $\det X(t) \neq 0$ for all $t \in \mathbb{R}$.

Example 4.1.2. Construct the transition matrix $X(t,s)$ of (4.1), using the transition matrix $U(t,s)$ of (4.3) and matrices $B_i, i \in \mathbb{Z}$.

Solution. We have found above that $X(t,s) = X(t)X^{-1}(s)$, where $X(t)$ is a fundamental matrix that could be defined by (4.5). Obviously, there exist several possible situations for t and s in $\mathbb{R}$. Let us consider only one of them. Namely, assume that $\theta_k \le \theta_p < t \le \theta_{p+1}$ and $\theta_{l-1} < s \le \theta_l \le \theta_{k-1}$. Then, using (4.5) one can find that

$$X(t,s) = X(t)X^{-1}(s) = U(t,\theta_p)(\mathcal{I} + B_p)U(\theta_p,\theta_{p-1})\ldots$$

$$\ldots U(\theta_{k+1},\theta_k)(\mathcal{I} + B_k)U(\theta_{k-1},t_0)X_0[U(t,\theta_l)(\mathcal{I} + B_l)^{-1}U(\theta_l,\theta_{l+1})\ldots$$
$$\ldots U(\theta_{k-1},\theta_{k-1})(\mathcal{I} + B_{k-1})^{-1}U(\theta_{k-1},0)X_0]^{-1} =$$
$$U(t,\theta_p)(\mathcal{I} + B_p)U(\theta_p,\theta_{p-1})\ldots U(\theta_{k+1},\theta_k)(\mathcal{I} + B_k)U(\theta_{k-1},t_0)X_0 \times$$
$$X_0^{-1}U(0,\theta_{k-1})(\mathcal{I} + B_{k-1})\ldots U(\theta_l,s) = U(t,\theta_p)(\mathcal{I} + B_p)U(\theta_p,\theta_{p-1})\ldots$$
$$\ldots(\mathcal{I} + B_l)U(\theta_l,s).$$

Thus, we have obtained that

$$X(t,s) = U(t,\theta_p)(\mathcal{I} + B_p)U(\theta_p,\theta_{p-1})\ldots(\mathcal{I} + B_l)U(\theta_l,s). \tag{4.6}$$

So, the transition matrix in this particular case is defined. We propose to the reader to derive the formula (4.6) directly, using the common procedure of extension of solutions of Sect. 2.3 and the condition $X(s,s) = \mathcal{I}$.

Example 4.1.3. Construct the fundamental matrix $X(t)$, $X(0) = \mathcal{I}$ of the system

$$\begin{aligned} x_1' &= -x_2, \\ x_2' &= x_1, \\ \Delta x_1|_{t=i} &= kx_1, \\ \Delta x_2|_{t=i} &= kx_2. \end{aligned} \tag{4.7}$$

Solution. First, consider the case $t \ge 0$. It is known that

$$U(t,s) = \begin{pmatrix} \cos(t-s) & -\sin(t-s) \\ \sin(t-s) & \cos(t-s) \end{pmatrix},$$

is a transition matrix of the system

$$\begin{aligned} x_1' &= -x_2, \\ x_2' &= x_1, \end{aligned} \tag{4.8}$$

for all $t, s \in \mathbb{R}$.

Using (4.6) one can find that

$$X(t,s) = \begin{pmatrix} \cos(t-\theta_p) & -\sin(t-\theta_p) \\ \sin(t-\theta_p) & \cos(t-\theta_p) \end{pmatrix} \begin{pmatrix} 1+k & 0 \\ 0 & 1+k \end{pmatrix} \times$$
$$\begin{pmatrix} \cos(\theta_p-\theta_{p-1}) & -\sin(\theta_p-\theta_{p-1}) \\ \sin(\theta_p-\theta_{p-1}) & \cos(\theta_p-\theta_{p-1}) \end{pmatrix} \begin{pmatrix} 1+k & 0 \\ 0 & 1+k \end{pmatrix} \times$$
$$\cdots \begin{pmatrix} \cos(\theta_k -) & -\sin(\theta_k -) \\ \sin(\theta_k -) & \cos(\theta_k -) \end{pmatrix}, \tag{4.9}$$

where $\theta_{k-1} < 0 \le \theta_k, \theta_p < t \le \theta_{p+1}$.

The matrix

$$\begin{pmatrix} 1+k & 0 \\ 0 & 1+k \end{pmatrix}$$

commutes with any another matrix. Moreover,

$$\begin{pmatrix} \cos\alpha & -\sin\alpha \\ \sin\alpha & \cos\alpha \end{pmatrix} \begin{pmatrix} \cos\beta & -\sin\beta \\ \sin\beta & \cos\beta \end{pmatrix} = \begin{pmatrix} \cos(\alpha+\beta) & -\sin(\alpha+\beta) \\ \sin(\alpha+\beta) & \cos(\alpha+\beta) \end{pmatrix},$$

for arbitrary $\alpha, \beta \in \mathbb{R}$. Consequently,

$$X(t) = \begin{pmatrix} \cos t & -\sin t \\ \sin t & \cos t \end{pmatrix} (1+k)^{i([0,t))}. \tag{4.10}$$

Similarly, one can find that if $t < 0$, then

$$X(t) = \begin{pmatrix} \cos t & -\sin t \\ \sin t & \cos t \end{pmatrix} (1+k)^{-i([t,0))}. \tag{4.11}$$

Finally, we have that

$$X(t) = \begin{cases} \begin{pmatrix} 1 & 0 \\ 0 & 1 \end{pmatrix}, & t = 0, \\ \begin{pmatrix} \cos t & -\sin t \\ \sin t & \cos t \end{pmatrix} (1+k)^{i([0,t))}, & t > 0 \\ \begin{pmatrix} \cos t & -\sin t \\ \sin t & \cos t \end{pmatrix} (1+k)^{-i([t,0))}, & t < 0. \end{cases} \tag{4.12}$$

Exercise 4.1.5. Apply (4.12)

(a) to verify that formula $X(t,s) = X(t)X^{-1}(s)$ is valid for all $t, s \in \mathbb{R}$;
(b) to analyze stability of the zero solution, varying the coefficient k.

Exercise 4.1.6. Consider a solution $x(t,0,x_0)$ of the following linear system:

$$
\begin{aligned}
x_1' &= -x_2,\\
x_2' &= x_1,\\
\Delta x_1|_{t=2\pi i} &= -x_1,\\
\Delta x_2|_{t=2\pi i} &= x_2.
\end{aligned}
\tag{4.13}
$$

Show that there are values of $x_0 \in \mathbb{R}^2$, such that the solution does not exist on $(-\infty, 0]$. Explain this result.

Example 4.1.4. [142] Consider the system

$$
\begin{aligned}
x' &= Ax,\\
\Delta x|_{t=\theta_i} &= Bx,
\end{aligned}
\tag{4.14}
$$

where constant matrices A, B are such that $AB = BA$. Let λ be an eigenvalue of matrix A, and x_0 be the correspond eigenvector. Prove that $x(t) = e^{\lambda(t-t_0)}(\mathcal{I}+B)^{i([t_0,t))}x_0, t \geq 0$, is a solution of (4.14), such that $x(t_0) = x_0$.

Solution. To verify that the proposed function is a solution, one should check that it satisfies the differential equation and equation of jumps. Let us begin with the differential equation. We have that

$$
x'(t) = \lambda e^{\lambda(t-t_0)}(\mathcal{I}+B)^{i([t_0,t))}x_0 = e^{\lambda(t-t_0)}(\mathcal{I}+B)^{i([t_0,t))}\lambda x_0 =
$$

$$
e^{\lambda(t-t_0)}(\mathcal{I}+B)^{i([t_0,t))}Ax_0 = Ae^{\lambda(t-t_0)}(\mathcal{I}+B)^{i([t_0,t))}x_0 = Ax(t).
$$

Now, fix $i \in \mathbb{Z}$. Then for the given function one can see that

$$
\begin{aligned}
\Delta x|_{t=\theta_i} &= x(\theta_i+) - x(\theta_i) = e^{\lambda(\theta_i - t_0+)}(\mathcal{I}+B)^{i([t_0,\theta_i+))}x_0 - e^{\lambda(\theta_i-t_0)}\\
&\qquad \times (\mathcal{I}+B)^{i([t_0,\theta_i))}x_0\\
&= [e^{\lambda(\theta_i-t_0+)}(\mathcal{I}+B) - e^{\lambda(\theta_i-t_0)}\mathcal{I}](\mathcal{I}+B)^{i([t_0,\theta_i))}x_0\\
&= [(\mathcal{I}+B)-\mathcal{I}]e^{\lambda(\theta_i-t_0)}(\mathcal{I}+B)^{i([t_0,\theta_i))}x_0 = Bx(\theta_i).
\end{aligned}
$$

Thus, $e^{\lambda(t-t_0)}(\mathcal{I}+B)^{i([t_0,t))}x_0$ is a solution. The initial condition can be verified easily.

Exercise 4.1.7. Use the result of the last example to determine a real valued solution of (4.14) if the eigenvalue is a complex number.

Exercise 4.1.8. Assume that the real parts of all eigenvalues of the matrix $A + \ln(\mathcal{I}+B)$ are negative, and there exists a positive number τ such that $i\tau \leq \theta_i \leq (i+1)\tau, i \in \mathbb{Z}$. Use the last two results to prove that there exist positive numbers N and ω, such that $||X(t,s)|| \leq N\mathrm{e}^{-\omega(t-s)}, t \geq s$, where $X(t,s)$ is the transition matrix of the system (4.14). Evaluate the numbers N and ω.

Example 4.1.5. Investigate asymptotic behavior of solutions of the system

$$\begin{aligned} x' &= ax, \\ \Delta x|_{t=\theta_i} &= bx, \end{aligned} \tag{4.15}$$

where $t, x \in \mathbb{R}, i \in \mathbb{Z}$, coefficients a and b are real numbers, $b \neq -1$, $|\theta_i| \to \infty$ as $|i| \to \infty$.

Solution. Since the equation is linear, $x(t, 0, x_0) = x(t, 0, 1)x_0$. So, it is sufficient to consider the behavior of $x_1(t) = x(t, 0, 1)$, the fundamental solution of (4.15). We shall consider the following cases.

(a) First, let us discuss the problem with $\theta_i = i\omega, i \in \mathbb{Z}$, where ω is a fixed positive real number. Applying (4.5), one can find that

$$x_1(t) = \begin{cases} \mathrm{e}^{at}(1+b)^{i([0,t))}, & t \geq 0, \\ \mathrm{e}^{at}(1+b)^{-i([t,0))}, & t < 0. \end{cases} \tag{4.16}$$

Then

$$|x_1(t)| = \begin{cases} \mathrm{e}^{(1-\{\frac{t}{\omega}\})\ln|1+b|}\mathrm{e}^{(a+\frac{\ln|1+b|}{\omega})t}, & t \geq 0, \\ \mathrm{e}^{-\{\frac{t}{\omega}\}\ln|1+b|}\mathrm{e}^{(a+\frac{\ln|1+b|}{\omega})t}, & t < 0. \end{cases} \tag{4.17}$$

where $\{t\} = t - [t]$. Let $\kappa = a + \frac{\ln|1+b|}{\omega}$. From (4.17) it follows that there are positive numbers m and M such that

$$m\mathrm{e}^{\kappa t} \leq |x_1(t)| \leq M\mathrm{e}^{\kappa t}, t \in \mathbb{R} \tag{4.18}$$

and

$$\kappa = \lim_{t\to\infty} \frac{1}{t} \ln|x_1(t)|. \tag{4.19}$$

That is, κ is an exponent of $x_1(t)$ and, consequently, it is the exponent of (4.15). Using (4.17) one can make the following conclusions:

(i) if $\kappa > 0$, then every nonzero solution $x(t)$ of (4.15) satisfies $|x(t)| \to \infty$ as $t \to \infty$, and $|x(t)| \to 0$ as $t \to -\infty$;
(ii) if $\kappa < 0$, then every nonzero solution $x(t)$ of (4.15) satisfies $|x(t)| \to 0$ as $t \to \infty$, and $|x(t)| \to \infty$ as $t \to -\infty$;
(iii) if $\kappa = 0$, then all solutions of (4.15) are functions bounded on $\mathbb{R}$.

(b) Assume that $\theta_{i+1} - \theta_i = \omega, i \in \mathbb{Z}$, where ω is a positive real number. Prove yourself that the solution $x_1(t) = x(t, 0, 1)$ of (4.15) and the number $\kappa = a + \frac{\ln|1+b|}{\omega}$ satisfy relations (4.18) and (4.19).
(c) Assume that the sequence $\theta_i, i \in \mathbb{Z}$, satisfies a more general condition than in (a) and (b). Namely, assume that the following limit:

$$\lim_{t-s\to\infty} \frac{i((s,t))}{t-s} = q \geq 0$$

exists.

It is obvious that if $x(t) = x(t,s,x_0), x(s) = x_0$ is a solution of (4.15), then $x(t) = x(t,s)x_0$, where

$$x(t,s) = \begin{cases} e^{a(t-s)}(1+b)^{i([s,t))}, & t \geq s, \\ e^{a(t-s)}(1+b)^{-i([t,s))}, & t < s. \end{cases} \tag{4.20}$$

is the transition "matrix" of (4.15).

So, it is sufficient to analyze $x(t,s)$ if one wants to investigate the asymptotic behavior of solutions of (4.15).

Fix $\epsilon > 0$ and set $\epsilon_1 = \frac{\epsilon}{\ln|1+b|}$. There exists a positive number $T(\epsilon_1)$ such that if $t - s \geq T(\epsilon_1)$, then

$$q - \epsilon < \frac{i([s,t])}{t-s} < q + \epsilon.$$

That is, each interval with the length $T(\epsilon_1)$ has at most $[(q+\epsilon)T(\epsilon_1)]$ points of θ. Consequently, there exist

$$M(\epsilon) = \sup_{0 \leq t-s \leq T} e^{a(t-s)}|1+b|^{i([s,t))}$$

and

$$m(\epsilon) = \inf_{0 \leq t-s \leq T} e^{a(t-s)}|1+b|^{i([s,t))}.$$

Hence,

$$m(\epsilon)e^{(\alpha-\epsilon_1)(t-s)} \leq |x(t,s)| \leq M(\epsilon)e^{(\alpha+\epsilon_1)(t-s)}, \ t \geq s, \tag{4.21}$$

where $\alpha = a + q\ln|1+b|$.

Exercise 4.1.9. Investigate asymptotic behavior of solutions of (4.15) if θ is a finite set.

Exercise 4.1.10. Solve the following problems.

1. Prove that (4.21) implies

$$\alpha = \lim_{t-s\to\infty} \frac{1}{t-s} \ln|x(t,s)|. \tag{4.22}$$

2. Prove analogs of (4.21) and (4.22) if $t - s < 0$.
3. Let the following relation be valid:

$$\sup_{s\in\mathbb{R}} \lim_{t-s\to\infty} \frac{i((s,t))}{t-s} = q \geq 0, \tag{4.23}$$

   and set $\alpha = a + q\ln|1+b|$.
   Prove that

$$\alpha = \lim_{t-s\to\infty} \frac{1}{t-s} \ln|x(t,s)|.$$

4. Assume that for some numbers $p \in \mathbb{N}, \omega \in \mathbb{R}, \omega > 0$, condition $(C6)$ of Chap. 3 is fulfilled. Applying the Poincaré criterion, find a sufficient condition for ω-periodicity of all solutions of (4.15).
5. Let the following system be given:

$$\begin{aligned} x' &= ax, \\ \Delta x|_{t=\theta_i} &= b_i x, \end{aligned} \tag{4.24}$$

where $t, x \in \mathbb{R}$, a and b_i are real coefficients. Assume that there exist limits

$$\lim_{t-s\to\infty} \frac{i([s,t])}{t-s} = q, \quad \lim_{i\to\infty} |1+b_i| = \beta.$$

Show that all solutions of the system tend to the zero as $t \to \infty$ if $a + q\ln\beta < 0$.

The adjoint system. Consider, besides (4.1), the linear system of impulsive differential equations

$$\begin{aligned} y' &= P(t)y, \\ \Delta y|_{t=\theta_i} &= Q_i y. \end{aligned} \tag{4.25}$$

where $(t, y) \in \mathbb{R} \times \mathbb{R}^n$, and the sequence $\theta_i, i \in \mathbb{Z}$, is the same as in (4.1). We suppose that entries of the matrix $P(t)$ belong to $\mathcal{PC}(\mathbb{R}, \theta)$, the real valued $n \times n$ matrices $Q_i, i \in \mathbb{Z}$, satisfy

$$\det(\mathcal{I} + Q_i) \neq 0. \tag{4.26}$$

One can easily see that all solutions of system (4.25) exist on $\mathbb{R}$ and are unique.

Definition 4.1.1. Systems (4.1) and (4.25) are mutually adjoint if any two solutions $x(t)$ and $y(t)$ of these equations satisfy

$$< x(t), y(t) >= c, \tag{4.27}$$

where $t \in \mathbb{R}$, $< \cdot, \cdot >$ is the scalar product and c is a real constant, which depends on these solutions.

Exercise 4.1.11. Prove that systems (4.1) and (4.25) are mutually adjoint if and only if any two solutions $x(t)$ and $y(t)$ of these systems satisfy conditions:

(a) $\frac{d<x(t),y(t)>}{dt} = 0$;
(b) $\Delta < x(t), y(t) > |_{t=\theta_i} \equiv < x(\theta_i+), y(\theta_i+) > - < x(\theta_i), y(\theta_i) >= 0$, $i \in \mathbb{Z}$.

Theorem 4.1.3. *Systems (4.1) and (4.25) are mutually adjoint if and only if* $P(t) = -A^T(t)$ *and* $Q_i = -(\mathcal{I} + B_i^T)^{-1} B_i^T$.

Proof. Sufficiency. It is easily seen that entries of matrix $-A^T(t)$ belong to $\mathcal{PC}(\mathbb{R}, \theta)$. Moreover, we have that $\mathcal{I} - (\mathcal{I} + B_i^T)^{-1} B_i^T = (\mathcal{I} + B_i^T)^{-1}$

$(\mathcal{I} + B_i^T - B_i^T) = (\mathcal{I} + B_i^T)^{-1}$ is not a singular matrix for an integer i. Consequently, each solution of the system exists on $\mathbb{R}$. Next, we will apply the results of the last exercise to prove the sufficiency. We have that

$$\frac{d < x(t), y(t) >}{dt} =< x'(t), y(t) > + < x(t), y'(t) >=< A(t)x(t), y(t) > +$$

$$< x(t), -A(t)^T y(t) >=< x(t), (A(t)^T - A(t)^T)y(t) >=< x(t), 0 >= 0.$$

Moreover, if $i \in \mathbb{Z}$, then

$$\Delta < x(t), y(t) > |_{t=\theta_i} =< x(\theta_i+), y(\theta_i+) > - < x(\theta_i), y(\theta_i) >=$$

$$< x(\theta_i+) - x(\theta_i), y(\theta_i+) > + < x(\theta_i), y(\theta_i+) - y(\theta_i) >=$$

$$< (B_i x(\theta_i), y(\theta_i+) > + < x(\theta_i), -(\mathcal{I} + B_i^T)^{-1} B_i^T y(\theta_i) >=$$

$$< x(\theta_i), (B_i^T(\mathcal{I} - (\mathcal{I} + B_i^T)^{-1} B_i^T)) - (\mathcal{I} + B_i^T)^{-1} B_i^T)y(\theta_i) >=< x(\theta_i), 0 >= 0.$$

The sufficiency is proved. □

Necessity. Assume that systems (4.1) and (4.25) are adjoint. Consider arbitrary solutions $x(t)$ and $y(t)$ of these systems. By Exercise 4.1.11 we have that $< x(t), y(t) >' = 0$ and $\Delta < x(t), y(t) > |_{t=\theta_i} = 0, i \in \mathbb{Z}$. Consequently, $< x, (A^T(t) + P(t))y >= 0, t \in \mathbb{R}$ and $< x, ((\mathcal{I} + B_i^T)^{-1} B_i^T + Q_i)y >= 0, i \in \mathbb{Z}$, for arbitrary $x, y \in \mathbb{R}^n$. The last two expressions imply that $P(t) = -A^T(t)$, and $Q_i = -(\mathcal{I} + B_i^T)^{-1} B_i^T$. Indeed, let us verify the second equation. Assume on the contrary, that $(\mathcal{I} + B_i^T)^{-1} B_i^T + Q_i \neq 0$, for some $i \in \mathbb{Z}$. Then one can find a vector $\bar{y} \in \mathbb{R}^n$ such that $((\mathcal{I} + B_i^T)^{-1} B_i^T + Q_i)\bar{y} \neq 0$. If $x = ((\mathcal{I} + B_i^T)^{-1} B_i^T + Q_i)\bar{y}$, then $< x, ((\mathcal{I} + B_i^T)^{-1} B_i^T + Q_i)\bar{y} >\neq 0$, and the last inequality contradicts the previous conclusion. Consequently, $(\mathcal{I} + B_i^T)^{-1} B_i^T + Q_i = 0$. The lemma is proved. □

Exercise 4.1.12. Prove that $X^T(t)^{-1}$ is a fundamental matrix of the adjoint system (4.25) if $X(t)$ is a fundamental matrix of (4.1).

Linear exponentially dichotomous systems. Fix a natural number $m, 0 < m < n$. Suppose that there exist m and $(n - m)$-dimensional hyperplanes $X_+(t)$ and $X_-(t)$ of $\mathbb{R}^n$ respectively such that if $x(t)$ is a solution of (4.1) and $x(t) \in X_+(t)$, then $\|x(t)\| \leq a_1 \|x(s)\| e^{-\gamma_1(t-s)}$, $-\infty < s \leq t < +\infty$ and, if $x(t) \in X_-(t)$, then $\|x(t)\| \geq a_2 \|x(s)\| e^{\gamma_2(t-s)}$, $-\infty < s \leq t < +\infty$. Here a_j, γ_j, $j = 1, 2$, are positive constants. Then (4.1) is said to be an *exponentially dichotomous* linear impulsive system.

Assume that the matrix $A(t)$ is bounded on $\mathbb{R}$. Let us show that by a piecewise-continuous Lyapunov transformation system (4.1) can be reduced to a box-diagonal system, i.e., a system splitting into two equations:

$$\frac{d\xi}{dt} = P_1(t)\xi, \ \Delta\xi|_{t=\theta_i} = Q_i^1\xi, \tag{4.28}$$

and

$$\frac{d\eta}{dt} = P_2(t)\eta, \ \Delta\eta|_{t=\theta_i} = Q_i^2\eta. \tag{4.29}$$

By applying the Gram–Schmidt process one can obtain a linear transformation $x(t) = U(t)y(t), y = (\xi, \eta)$, which takes system (4.3) to the form

$$\frac{d\xi}{dt} = P_1(t)\xi, \ \frac{d\eta}{dt} = P_2(t)\eta, \tag{4.30}$$

on each interval of continuity $(\theta_i, \theta_{i+1}), i \in \mathbb{Z}$. Matrix $U(t)$ is continuous on the union of intervals $(\theta_i, \theta_{i+1}), i \in \mathbb{Z}$, and is uniformly bounded on the set together with matrices $dU(t)/dt, U^{-1}(t)$. For each θ_i one can continue $U(\theta_i) = U(\theta_i-)$. That is, $U(t)$ is a Lyapunov piecewise continuous matrix [142]. Since solutions of impulsive differential equations are left-continuous functions, $x(\theta_i) = U(\theta_i)y(\theta_i)$ for all i. Let $X(t)$ be the fundamental matrix of (4.1), which is used to define $U(t), U(t) = X(t)S(t), S(t) = diag(S_+, S_-)$. Then $U(\theta_i+) = X(\theta_i+)S(\theta_i+) = (\mathcal{I} + B_i)X(\theta_i)S(\theta_i+)$ and $U(\theta_i) = X(\theta_i)S(\theta_i)$. Subtract from the first equality the second one to obtain $\Delta U|_{t=\theta_i} = (\mathcal{I} + B_i)X(\theta_i)\Delta S + B_i U(\theta_i)$ or

$$U^{-1}(\theta_i+)[B_i U(\theta_i) - \Delta U] = U^{-1}(\theta_i+)(\mathcal{I} + B_i)X(\theta_i)\Delta S.$$

It implies that

$$U^{-1}(\theta_i+)[B_i U(\theta_i) - \Delta U] = -S^{-1}(\theta_i+)\Delta S.$$

That is, $Q_i = U^{-1}(\theta_i+)[B_i U(\theta_i) - \Delta U], i \in \mathbb{Z}$, are box-diagonal matrices. Consequently, using the transformation $x = U(t)y$ in the equation $\Delta x|_{t=\theta_i} = B_i x$, we obtain that $\Delta y|_{t=\theta_i} = Q_i y$. Denoting $Q_i = diag(Q_i^1, Q_i^2)$, we arrive to the system of equations (4.28) and (4.29). Next, one can obtain that there exist positive constants K, γ such that

$$\|X_1(t,s)\| \le K\exp(-\gamma(t-s)), \ t \ge s, \tag{4.31}$$

and

$$\|X_2(t,s)\| \le K\exp(\gamma(t-s)), \ t \le s, \tag{4.32}$$

where $X_1(t,s)$ and $X_2(t,s)$ are transition matrices of (4.28) and (4.29) respectively.

4.2 Linear Nonhomogeneous Systems

Consider the following system:

$$\begin{aligned} y' &= A(t)y + f(t), \\ \Delta y|_{t=\theta_i} &= B_i y + J_i, \end{aligned} \tag{4.33}$$

where $(t, x) \in \mathbb{R} \times \mathbb{R}^n$, an infinite sequence θ_i satisfies $|\theta_i| \to \infty$ as $|i| \to \infty$. It is assumed that there exists a positive constant θ such that $\theta_{i+1} - \theta_i \leq \theta$. Moreover, real valued entries of the matrix $A(t)$ are from $\mathcal{PC}(\mathbb{R}, \theta)$, bounded on $\mathbb{R}$, and real valued $n \times n$ matrices $B_i, i \in \mathbb{Z}$, satisfy (4.2). Coordinates of the vector-function $f(t) : \mathbb{R} \to \mathbb{R}^n$ belong to $\mathcal{PC}(\mathbb{R}, \theta)$, and $J_i, i \in \mathbb{Z}$, is a sequence of vectors from $\mathbb{R}^n$. We assume that

$$\sup_t ||f|| + \sup_i ||J|| = \bar{M} < \infty. \tag{4.34}$$

Repeating identically the proof of Theorem 4.1.1, one can check that the following assertion is valid.

Theorem 4.2.1. *Every solution $x(t) = x(t, t_0, x_0), (t_0, x_0) \in \mathbb{R} \times \mathbb{R}^n$, of (4.33) is unique and continuable on $\mathbb{R}$.*

The general solution of (4.33). Fix $(t_0, x_0) \in \mathbb{R} \times \mathbb{R}^n$. Let $X(t), X(t_0) = \mathcal{I}$, be a fundamental matrix of (4.1), associated with (4.33). Let us apply to system (4.33) the transformation $y = X(t)z$, where $z \in \mathbb{R}^n$ is a new variable, depending on t. We have that $X(t)z' + X'(t)z = A(t)X(t)z + f(t)$ or

$$z'(t) = X^{-1}(t)f(t). \tag{4.35}$$

If $t = \theta_i$, for some $i \in \mathbb{Z}$, then the substitution implies that

$$X(\theta_i+)z(\theta_i+) - X(\theta_i)z(\theta_i) = B_i X(\theta_i)z(\theta_i) + J_i,$$

and

$$X(\theta_i+)(z(\theta_i+) - z(\theta_i)) + (X(\theta_i+) - X(\theta_i))z(\theta_i) = B_i X(\theta_i)z(\theta_i) + J_i.$$

Since $X(\theta_i+) - X(\theta_i) = B_i X(\theta_i)$, the last formula yields that

$$\Delta z|_{t=\theta_i} = X^{-1}(\theta_i+)J_i, i \in \mathbb{Z}. \tag{4.36}$$

Thus, combining (4.35) with (4.36), $y(t)$ is a solution of (4.33) if and only if $z(t) = X^{-1}(t)y(t)$ is a solution of the system

$$\begin{aligned} z'(t) &= X^{-1}(t)f(t) \\ \Delta z|_{t=\theta_i} &= X^{-1}(\theta_i+)J_i. \end{aligned} \tag{4.37}$$

Exercise 4.2.1. Prove that the coordinates of $X^{-1}(t)f(t)$ belong to $\mathcal{PC}(\mathbb{R}, \theta)$.

A solution $z(t) = z(t, t_0, z_0)$ of (4.37) can be easily found, similarly to the results of Sect. 2.7,

$$z(t) = \begin{cases} z_0 + \int_{t_0}^t X^{-1}(s)f(s)ds + \sum_{t_0 \leq \theta_i < t} X^{-1}(\theta_i+)J_i, & t \geq t_0, \\ z_0 + \int_{t_0}^t X^{-1}(s)f(s)ds - \sum_{t \leq \theta_i < t_0} X^{-1}(\theta_i+)J_i, & t < t_0. \end{cases} \tag{4.38}$$

Now, taking into account that $y_0 = X(t_0)z_0 = z_0$, and making the inverse substitution one can derive that

$$x(t,t_0,x_0) = \begin{cases} X(t)x_0 + \int\limits_{t_0}^{t} X(t)X^{-1}(s)f(s)ds + \sum_{t_0 \le \theta_i < t} X(t)X^{-1}(\theta_i+)J_i, & t \ge t_0, \\ X(t)x_0 + \int\limits_{t_0}^{t} X(t)X^{-1}(s)f(s)ds - \sum_{t \le \theta_i < t_0} X(t)X^{-1}(\theta_i+)J_i, & t < t_0. \end{cases}$$

If the fundamental matrix is not necessarily the unit matrix at $t = t_0$, then one can write

$$x(t,t_0,x_0) = \begin{cases} X(t,t_0)x_0 + \int\limits_{t_0}^{t} X(t,s)f(s)ds + \sum\limits_{t_0 \le \theta_i < t} X(t,\theta_i+)J_i, & t \ge t_0, \\ X(t,t_0)x_0 + \int\limits_{t_0}^{t} X(t,s)f(s)ds - \sum\limits_{t \le \theta_i < t_0} X(t,\theta_i+)J_i, & t < t_0. \end{cases} \tag{4.39}$$

Last two formulas define the general solution of the linear nonhomogeneous system (4.33).

Example 4.2.1. Consider the following system:

$$\begin{aligned} y' &= ay + f(t), \\ \Delta y|_{t=\theta_i} &= by + w_i, \end{aligned} \tag{4.40}$$

where $t, y \in \mathbb{R}$, a and $b, b \neq -1$, are real constants, $f \in \mathcal{PC}(\mathbb{R},\theta)$, $w_i \in \mathbb{R}, i \in \mathbb{Z}$. Assume that

$$\sup_t |f| + \sup_i |w_i| = M < \infty.$$

We suppose that θ is a B-sequence, and there exists the limit

$$\lim_{t-s\to\infty} \frac{i([s,t])}{t-s} = q \ge 0.$$

Write $\alpha = a + q \ln|1+b|$, and assume that $\alpha \neq 0$.

Let us show that (4.40) admits a unique solution $y_0(t)$ bounded on $\mathbb{R}$. Consider $\alpha > 0$. Results of Example 4.1.5, (c) imply that there exist positive constants β, K such that

$$|x(t,s)| \le K e^{\beta(t-s)}, t \le s, \tag{4.41}$$
$$|x(t,s)| \ge K^{-1} e^{\beta(t-s)}, t \ge s, \tag{4.42}$$

where $x(t,s)$ is the fundamental solution of (4.15). The solution $y(t) = y(t,0,y_0)$ of (4.40) has the form

$$y(t) = x(t,0)y_0 + \int\limits_0^t x(t,s)f(s)ds + \sum_{0 \le \theta_i < t} x(t,\theta_i+)w_i$$

or

$$y(t) = x(t,0)[y_0 + \int_0^t x(0,s)f(s)ds + \sum_{0 \le \theta_i < t} x(0,\theta_i+)w_i]. \tag{4.43}$$

The last formula and condition (4.42) imply that the solution is bounded only if

$$y_0 = -\int_0^\infty x(0,s)f(s)ds - \sum_{0 \le \theta_i < \infty} x(0,\theta_i+)w_i. \tag{4.44}$$

Let us show that the integral and sum in the last formula are convergent. Using (4.41) we have that

$$|\int_0^\infty x(0,s)f(s)ds| \le \frac{MK}{\beta} < \infty.$$

Further, one can obtain that

$$|\sum_{0 \le \theta_i < \infty} x(0,\theta_i+)w_i| \le MK \sum_{0 \le \theta_i < \infty} e^{-\beta\theta_i}.$$

Fix $\epsilon > 0$. Then $i([s,t]) \le (q+\epsilon)(t-s)$, if $t-s \ge T(\epsilon)$ for some positive $T(\epsilon)$.

Thus, every interval of the length $T(\epsilon)$ consists of not more than $[(q+\epsilon)(t-s)]$ elements of θ. Hence,

$$\sum_{0 \le \theta_i < \infty} e^{-\beta\theta_i} \le \sum_{i=0}^\infty (q+\epsilon)T(\epsilon)e^{-i\beta T(\epsilon)} = (q+\epsilon)T(\epsilon)\frac{1}{1-e^{\beta T(\epsilon)}}.$$

Thus, the integral and sum are convergent. Using the value of y_0 in (4.43) one can find that

$$y_0(t) = -\int_t^\infty x(t,s)f(s)ds - \sum_{t \le \theta_i < \infty} x(t,\theta_i+)w_i. \tag{4.45}$$

If $\epsilon > 0$ is fixed, then one can obtain that

$$|y_0(t)| \le MK[\frac{1}{\beta} + (q+\epsilon)T(\epsilon)\frac{1}{1-e^{\beta T(\epsilon)}}] < \infty, \tag{4.46}$$

for all $t \in \mathbb{R}$.

Exercise 4.2.2. Solve the following problems.

1. Verify that (4.46) is valid;
2. Show, directly, that $y_0(t)$ is a solution of (4.40);
3. Prove that $y_0(t)$ is a unique solution of (4.40) bounded on $\mathbb{R}$.

Exercise 4.2.3. Assume that $\alpha < 0$ in the last example. Prove that

$$y_0(t) = \int_{-\infty}^{t} x(t,s)f(s)ds + \sum_{t<\theta_i} x(t,\theta_i+)w_i$$

is a unique solution of (4.40) bounded on $\mathbb{R}$.

In the rest of this section, we shall develop the results of the last example and exercise to the most general case. Consider system (4.33) again assuming that associated homogeneous system (4.1) is exponentially dichotomous. That is, there exists a linear transformation $U(t)$, which reduces (4.1) to (4.28) and (4.29). Apply the substitution in (4.33) and obtain the following system of linear nonhomogeneous impulsive differential equations:

$$\begin{aligned} \frac{d\xi}{dt} &= P_1(t)\xi + f_1(t), \\ \Delta\xi|_{t=\theta_i} &= Q_i^1\xi + J_i^1, \end{aligned} \tag{4.47}$$

$$\begin{aligned} \frac{d\eta}{dt} &= P_2(t)\eta + f_2(t), \\ \Delta\eta|_{t=\theta_i} &= Q_i^2\eta + J_i^2, \end{aligned} \tag{4.48}$$

where $x = U(t)y, y = (\xi,\eta)$, matrices P_1, P_2, Q_i^1, Q_i^2, are defined in (4.28) and (4.29), and $(f_1, f_2) = U^{-1}(t)f, (J_i^1, J_i^2) = U^{-1}(\theta_i+)J_i$. On the basis of the assumptions made above, we have that

$$\sup_t ||(f_1, f_2)|| + \sup_i ||(J_i^1, J_i^2)|| = M < \infty.$$

Theorem 4.2.2. *If the associated linear homogeneous system (4.1) is exponentially dichotomous and (4.34) is valid, then there exists a unique solution of (4.33) bounded on $\mathbb{R}$.*

Proof. It is sufficient to show that functions

$$\xi(t) = \int_{-\infty}^{t} X_1(t,s)f_1(s)ds + \sum_{-\infty<\theta_i<t} X_1(t,\theta_i+)J_i^1, \tag{4.49}$$

and

$$\eta(t) = -\int_{t}^{\infty} X_2(t,s)f_2(s)ds - \sum_{t\le\theta_i<\infty} X_2(t,\theta_i+)J_i^2. \tag{4.50}$$

are components of a unique bounded solution of the system of equations (4.47) and (4.48).

(i). We start with (4.49). First of all one has that

$$||\xi(t)|| \leq \int_{-\infty}^{t} Ke^{-\gamma(t-s)}Mds + \sum_{-\infty<\theta_i<t} Ke^{-\gamma(t-\theta_i)}M \leq KM[\frac{1}{\gamma} + \frac{1}{1-e^{-\gamma\theta}}].$$

That is, both the integral and infinite sum are convergent uniformly for all t, and the function ξ is bounded on $\mathbb{R}$. Differentiating, we have that

$$\xi'(t) = X_1(t,t)f_1(t) + \int_{-\infty}^{t} X_1'(t,s)f_1(s)ds + \sum_{-\infty<\theta_i<t} X_1'(t,\theta_i+)J_i^1 =$$

$$f_1(t) + \int_{-\infty}^{t} P_1(t)X_1(t,s)f_1(s)ds + \sum_{-\infty<\theta_i<t} P_1(t)X_1(t,\theta_i+)J_i^1 = P_1(t)\xi(t) + f_1(t).$$

Fix $j \in \mathbb{Z}$ and verify that

$$\Delta\xi|_{t=\theta_j} = \xi(\theta_j+) - \xi(\theta_j) = \int_{-\infty}^{\theta_j+} X_1(\theta_j+,s)f_1(s)ds + \sum_{-\infty<\theta_i<\theta_j+} X_1(\theta_j+,\theta_i+)J_i^1 -$$

$$\int_{-\infty}^{\theta_j} X_1(\theta_j,s)f_1(s)ds - \sum_{-\infty<\theta_i<\theta_j} X_1(\theta_j,\theta_i+)J_i^1 = \int_{-\infty}^{\theta_j} Q_i^1 X_1(\theta_j,s)f_1(s)ds +$$

$$\sum_{-\infty<\theta_i<\theta_j} Q_i^1 X_1(\theta_j,\theta_i+)J_i^1 + J_j^1 = Q_j^1\xi(\theta_i) + J_j^1.$$

Thus, ξ is a solution of (4.49). Assume that the equation has another solution $\phi(t)$ bounded on $\mathbb{R}$. Then, the difference $\xi - \phi$ is a bounded solution of the system (4.28). The general solution of the equation is $X_1(t,t_0)\xi_0$. The inequality $1 = ||\mathcal{I}|| \leq ||X_1(t,s)||||X_1(s,t)||$ implies that $||X_1(s,t)|| \geq K^{-1}e^{\gamma(t-s)}, s \leq t$. That is, $||X_1(t,t_0)|| \to \infty$, as $t \to -\infty$, and (4.28) admits a unique bounded solution, $\xi \equiv 0$. Hence, $\phi = \xi$.

(ii). Consider the function η, now. Evaluating

$$||\eta(t)|| \leq \int_{t}^{\infty} Ke^{\gamma(t-s)}Mds + \sum_{t\leq\theta_i<\infty} Ke^{\gamma(t-\theta_i)}M \leq KM[\frac{1}{\gamma} + \frac{1}{1-e^{-\gamma\theta}}],$$

we obtain that η is a bounded function. To show that it is a solution of (4.50), substitute the function in the system. We have that

$$\eta'(t) = X_2(t,t)f_2(t) - \int_t^{\infty} X_2'(t,s)f_2(s)ds - \sum_{t\le\theta_i<\infty} X_2'(t,\theta_i+)J_i^2 =$$

$$f_2(t) - \int_t^{\infty} P_2(t)X_2(t,s)f_2(s)ds - \sum_{t\le\theta_i<\infty} P_2(t)X_2(t,\theta_i+)J_i^2 = P_2(t)\eta(t) + f_2(t).$$

Fix $j \in \mathbb{Z}$, and obtain that

$$\Delta\eta|_{t=\theta_j} = -\int_{\theta_j+}^{\infty} X_2(\theta_j+,s)f_2(s)ds - \sum_{\theta_j+\le\theta_i<\infty} X_2(\theta_j+,\theta_i+)J_i^2 +$$

$$\int_{\theta_j}^{\infty} X_2(\theta_j,s)f_2(s)ds + \sum_{\theta_j\le\theta_i<\infty} X_2(\theta_j,\theta_i+)J_i^2 = -\int_{\theta_j}^{\infty} Q_i^2 X_2(\theta_j,s)f_2(s)ds -$$

$$\sum_{\theta_j\le\theta_i<\infty} Q_i^2 X_2(\theta_j,\theta_i+)J_i^2 - (-J_j) = Q_i^2\eta(\theta_i) + J_j.$$

That is, η is a solution of (4.50). Uniqueness of this solution can be verified in the same way as that of $\xi(t)$. The theorem is proved. □

4.3 Linear Periodic Systems

In this section the basic information on the important subject, existence of periodic solutions and their stability is discussed. We investigate linear homogeneous and nonhomogeneous systems with periodic coefficients. Consider (4.1) assuming this time that it is (ω, p)-periodic system. That is, $A(t)$ is an ω-periodic matrix-function, B_i is a p-periodic sequence of matrices, and $\theta_{i+p} = \theta_i + \omega$ for all $i \in \mathbb{Z}$, that is the sequence θ has the p-property. If $\det(\mathcal{I} + B_i) \neq 0, i \in \mathbb{Z}$, then, by results of Sect. 4.1, there exists a fundamental matrix $X(t)$ of (4.1).

Exercise 4.3.1. Prove the following assertions.

1. If sequence θ has the p-property then for arbitrary $a \in \mathbb{R}$ there exist p numbers, $a \le \xi_1 < \xi_2 < \ldots < \xi_p < a + \omega$, such that for each $\theta_i \in \theta$ one can find uniquely an integer k and a number ξ_j, which satisfy $\theta_i = \xi_j + k\omega$;
2. The matrix $X(t+\omega)$ is also a fundamental matrix of (4.1) and

$$X(t+\omega) = X(t)X(\omega), \tag{4.51}$$

for all $t \in \mathbb{R}$;

3. It is true that

$$X(t+\omega)X^{-1}(s+\omega) = X(t)X^{-1}(s), \tag{4.52}$$

for all $t, s \in \mathbb{R}$.

If $X(t)$ is a fundamental matrix with $X(0) = \mathcal{I}$, then $X(\omega)$ is called a *monodromy matrix* and the eigenvalues of the monodromy matrix are *multipliers*. Denote the multipliers by $\rho_i, i = 1, 2, \ldots, n$. The role of multipliers for linear periodic systems is identical to one of eigenvalues for linear systems with constant coefficients. To emphasize the role we can introduce the special numbers, *exponents*, which are equal to $\lambda_i = \frac{1}{\omega}\mathrm{L}n\rho_i, i = 1, 2, \ldots, n$.

Thus, multipliers are solutions of the equation

$$\det(X(\omega) - \rho\mathcal{I}) = 0, \tag{4.53}$$

and exponents are solutions of the equation

$$\det(\frac{1}{\omega}\mathrm{L}n\, X(\omega) - \lambda\mathcal{I}) = 0. \tag{4.54}$$

Theorem 4.3.1. *A number ρ is a multiplier of (4.1) if and only if there exists a solution $x(t)$ of the system such that $x(t+\omega) = \rho x(t)$.*

The proof of the last theorem replicates that of the similar theorem for ordinary differential equations [59].

Exercise 4.3.2. Prove that (ω, p)-periodic system (4.1) has a periodic solution with period $k\omega, k = 1, 2. \ldots,$ if and only if the k-th power of a multiplier equals to one.

Exercise 4.3.3. Use the result of Exercise 4.1.12 to prove the following assertion.

Theorem 4.3.2. *[142] System (4.1) has $r, 1 \le r \le n$, linearly independent ω-periodic solutions if and only if the adjoint system*

$$\begin{aligned} y' &= -A^T(t)y, \\ \Delta y|_{t=\theta_i} &= -(\mathcal{I} + B_i^T)^{-1}B_i^T y \end{aligned} \tag{4.55}$$

has $r, 1 \le r \le n$, linearly independent ω-periodic solutions.

Suppose systems (4.1) and (4.55) have $r, 1 \le r \le n$, linearly independent ω-periodic solutions. Denote those of (4.55) as $\psi_j(t), j = 1, 2, \ldots, r$. Consider the problem of existence of periodic solutions of the following (ω, p)-periodic non-homogeneous system:

$$\begin{aligned} y' &= A(t)y + f(t), \\ \Delta x|_{t=\theta_i} &= B_i y + J_i. \end{aligned} \tag{4.56}$$

In what follows, we denote by $\mathcal{PC}_\omega(\theta) \subset \mathcal{PC}_p$ the set of all functions of fixed period ω and with sequence of discontinuity moments θ.

Exercise 4.3.4. Prove that the following assertion is valid.

Theorem 4.3.3. *[142] Assume that homogeneous system (4.1) has $r, 1 \leq r \leq n$, linearly independent ω-periodic solutions. System (4.56) admits an ω-periodic solution if and only if*

$$\int_0^\omega < \psi_j(t), f(t) > dt + \sum_{i=1}^{p} < \psi_j(\theta_i), J_i >= 0, \tag{4.57}$$

for all $j = 1, 2, \ldots, r$. In this case, system (4.56) has r-parametric family of ω-periodic solutions.

Let us introduce matrices $P = \frac{1}{\omega}\mathrm{Ln}X(\omega)$ and $F(t) = X(t)e^{-Pt}$. By applying (4.51), we find that

$$F(t+\omega) = X(t+\omega)e^{-P(t+\omega)} = X(t)X(\omega)e^{-P(t+\omega)} = F(t).$$

That is, $F(t) \in \mathcal{PC}_\omega(\theta)$.

Exercise 4.3.5. Prove that the following assertions are valid.

1. $F \in \mathcal{PC}^1(\mathbb{R}, \theta)$.
2. $|\det F(t)| \geq m$, for some positive number m and all $t \in \mathbb{R}$.
3. $F^{-1}(t), t \in \mathbb{R}$, exists, and its entries are from $\mathcal{PC}_\omega(\theta)$.

The following result is a basic one for the Floquet theory.

Theorem 4.3.4. *The substitution $x = F(t)y$ transforms (4.1) to the system with constant coefficients,*

$$\frac{dy}{dt} = Py. \tag{4.58}$$

Proof. First, we have that

$$F(t)y' + F'(t)y = X(t)e^{-Pt}y' + [X'(t)e^{-Pt} + X(t)(-P)e^{-Pt}]y = A(t)X(t)e^{-Pt}y$$

or $X(t)e^{-Pt}y' = X(t)Pe^{-Pt}y$. Cancellations in both parts of the last equality give us

$$y' = Py.$$

Moreover,

$$F(\theta_i+)y(\theta_i+) - F(\theta_i)y(\theta_i) = B_iF(\theta_i)y(\theta_i),$$

or

$$F(\theta_i+)\Delta y|_{t=\theta_i} + (F(\theta_i+) - F(\theta_i))y(\theta_i) = B_iF(\theta_i)y(\theta_i),$$

for a fixed $i \in \mathbb{Z}$.

Since $F(\theta_i+) - F(\theta_i) = B_i F(\theta_i)$, the last formula yields that

$$\Delta y|_{t=\theta_i} = 0.$$

That is, $y(t)$ satisfies (4.58). The theorem is proved. □

Exercise 4.3.6. Using the results of Exercise 4.3.5 prove that the following assertions are valid.

1. System (4.1) has a bounded solution if and only if (4.58) does.
2. System (4.1) is uniformly stable if and only if (4.58) is.
3. System (4.1) is uniformly asymptotically stable if and only if (4.58) is.
4. System (4.1) is unstable if and only if (4.58) is.
5. If system (4.1) is unstable then it has an unbounded solution.
6. If system (4.1) is stable then it is uniformly stable.
7. If system (4.1) is asymptotically stable then it is uniform asymptotically stable.

Exercise 4.3.7. Prove that the exponents $\lambda_i = \frac{1}{\omega}\mathrm{L}n\rho_i, i = 1,\ldots,n,$ are eigenvalues of the matrix P.

Exercise 4.3.8. Using Exercises 4.3.6 and 4.3.7 prove that the following theorem is valid.

Theorem 4.3.5. *[142] The periodic system (4.1) is:*

(i) uniformly stable if and only if all multipliers satisfy $|\rho_i| \le 1, i = 1,\ldots,n,$ and Jordan cells of the monodromy matrix, which correspond to multipliers with unit absolute values, have order one;
(ii) uniformly asymptotically stable if and only if all multipliers lie inside of the unite circle of the complex plane;
(iii) unstable if there exists a multiplier with absolute value larger than one.

Consider the following linear nonhomogeneous system

$$\begin{aligned} \frac{dx}{dt} &= A(t)x + f(t), \\ \Delta x|_{t=t_i} &= B_i x + I_i, \end{aligned} \tag{4.59}$$

where $A(t), f(t) \in \mathcal{PC}_\omega(\theta), B_i, I_i, i \in \mathbb{Z}$, are p-periodic sequences. That is, (4.59) is an (ω, p)-periodic system. Let (4.1) be the linear homogeneous system associated with (4.59). Denote by $X(t,s)$ the transition matrix of this system, and let $\lambda_i, i = 1,\ldots,n$, be the exponents of system (4.1).

Theorem 4.3.6. *If the real parts of the exponents $\lambda_i, i = 1,\ldots,n$, do not vanish, then (4.59) has a unique ω-periodic solution, which is uniformly asymptotically stable as soon as all of the exponents have negative real parts.*

Proof. Apply the Floquet transformation $x = F(t)y$ to (4.59). Then one can easily find that

$$\begin{aligned} \frac{dy}{dt} &= Py + F^{-1}(t)f(t), \\ \Delta y|_{t=\theta_i} &= F^{-1}(\theta_i+)I_i. \end{aligned} \quad (4.60)$$

Exponents $\lambda_i, i = 1, \ldots, n$, are eigenvalues of the matrix P. Therefore, there is a constant nonsingular matrix S such that the transformation $x = F(t)Sz$ reduces (4.59) to

$$\begin{aligned} \frac{dz}{dt} &= \Lambda z + g(t), \\ \Delta z|_{t=\theta_i} &= V_i, \end{aligned} \quad (4.61)$$

where $\Lambda = \text{diag}(\Lambda_+, \Lambda_-)$ is a constant box-diagonal matrix, $\text{Re}\lambda_j(\Lambda_+) < 0, j = 1, \ldots, m, \text{Re}\lambda_j(\Lambda_+) > 0, j = m+1, \ldots, n$, the integer m satisfies $0 \leq m \leq n$, and

$$\Lambda = S^{-1}F^{-1}(t)[A(t) - \frac{dF(t)}{dt}F^{-1}(t)]F(t)S,$$
$$g(t) = S^{-1}F^{-1}(t)f(t), \quad V_i = S^{-1}F^{-1}(\theta_i+)I_i.$$

One can easily see that (4.61) is an (ω, p)-periodic system. Set $g = (g_+, g_-)$, $V_i = (V_i^+, V_i^-), z = (z^+, z^-)$, such that system (4.61) has the form

$$\begin{aligned} \frac{dz^+}{dt} &= \Lambda_+ z^+ + g_+(t), \\ \Delta z^+|_{t=\theta_i} &= V_i^+, \end{aligned} \quad (4.62)$$

$$\begin{aligned} \frac{dz^-+}{dt} &= \Lambda_- z^- + g_+(t), \\ \Delta z^-|_{t=\theta_i} &= V_i^-, \end{aligned} \quad (4.63)$$

If $\gamma = \min_{1\leq j\leq n} |\text{Re}\lambda_j(\Lambda_\cdot)| + \epsilon$, where ϵ is a positive number, then there exists a number $K = K(\epsilon), K > 1$, such that $||e^{\Lambda_+ t}|| \leq Ke^{-\gamma t}, t \geq 0$, and $||e^{\Lambda_- t}|| \leq Ke^{\gamma t}, t \leq 0$. By Theorem 4.2.2, there is a unique bounded solution $z(t) = (z^+, z^-)$ of the system,

$$z^+(t) = \int_{-\infty}^{t} e^{\Lambda_+(t-s)}g_+(s)ds + \sum_{-\infty<\theta_i<t} e^{\Lambda_+(t-\theta_i)}V_i^+, \quad (4.64)$$

$$z^-(t) = -\int_{t}^{\infty} e^{\Lambda_-(t-s)}g_-(s)ds - \sum_{t\leq\theta_i<\infty} e^{\Lambda_-(t-\theta_i)}V_i^-. \quad (4.65)$$

So, it is sufficient to check that $z(t)$ is ω-periodic. Consider the question only for z^+, as for z^- the verification is very similar. We have that

$$z^+(t+\omega) = \int_{-\infty}^{t+\omega} e^{\Lambda_+(t+\omega-s)} g_+(s)ds + \sum_{-\infty<\theta_i<t+\omega} e^{\Lambda_+(t+\omega-\theta_i)} V_i^+ =$$

$$\int_{-\infty}^{t} e^{\Lambda_+(t+\omega-(s+\omega))} g_+(s+\omega)ds + \sum_{-\infty<\theta_i<t} e^{\Lambda_+(t+\omega-(\theta_i+\omega))} V_{i+p}^+ =$$

$$\int_{-\infty}^{t} e^{\Lambda_+(t-s)} g_+(s)ds + \sum_{-\infty<\theta_i<t} e^{\Lambda_+(t-\theta_i)} V_i^+ = z^+(t).$$

The theorem is proved. □

Next, let us consider the periodic systems applying the *Green's function* concept. Let (4.59) be again an (ω, p)-periodic system, and $X(t), X(0) = \mathcal{I}$, be the fundamental matrix of the associated homogeneous system (4.1). Without loss of generality we assume that $\theta_i \neq 0, i \in \mathbb{Z}$. Hence, $\theta_i \neq \omega, i \in \mathbb{Z}$. In what follows, we assume that $\det(\mathcal{I} - X(\omega)) \neq 0$ and introduce a Green's function

$$G(t,s) = \begin{cases} X(t)[\mathcal{I} - X(\omega)]^{-1}X^{-1}(s), & 0 \le s < t \le \omega, \\ X(t+\omega)[\mathcal{I} - X(\omega)]^{-1}X^{-1}(s), & 0 \le t \le s \le \omega. \end{cases} \tag{4.66}$$

Theorem 4.3.7. *The following properties are valid.*

1. $G_t'(t,s) = A(t)G(t,s), t \neq s;$
2. $\Delta G(t,s)|_{t=\theta_j} = G(\theta_j+,s) - G(\theta_j,s) = B_jG(\theta_j,s), s \neq \theta_j, j \in \mathbb{Z};$
3. $G(s+,s) - G(s,s) = \mathcal{I}, s \neq \theta_i, i = 1,2,\ldots,p;$
4. $G(0,s) - G(\omega,s) = 0, s \in [0,\omega);$
5. $G(0,\omega) - G(\omega,\omega) = \mathcal{I}.$

Exercise 4.3.9. Prove the last theorem.

Exercise 4.3.10. Using the last theorem, verify that the function

$$x(t) = \int_0^\omega G(t,s)f(s)ds + \sum_{0<\theta_i<\omega} G(t,\theta_i+)I_i \tag{4.67}$$

is an ω-periodic solution of (4.59).

Theorem 4.3.8. *If the associated homogeneous system (4.1) does not admit a nontrivial ω-periodic solution, then (4.59) has a unique ω-periodic solution equals (4.67).*

Proof. The solution $x(t)$ Of the initial value problem (4.59) and $x(t_0) = x_0$ is equal to

$$x(t) = X(t)x_0 + \int_0^t X(t)X^{-1}(s)f(s)ds + \sum_{0<\theta_i<t} X(t)X^{-1}(\theta_i+)I_i.$$

The Poincaré criterion implies that $x(t)$ is a unique ω-periodic solution of (4.59) if and only if x_0 satisfies the equation

$$[\mathcal{I} - X(\omega)]x_0 = \int_0^\omega X(\omega)X^{-1}(s)f(s)ds + \sum_{0<\theta_i<\omega} X(\omega)X^{-1}(\theta_i+)I_i$$

uniquely. That is,

$$x_0 = [\mathcal{I} - X(\omega)]^{-1}[\int_0^\omega X(\omega)X^{-1}(s)f(s)ds + \sum_{0<\theta_i<\omega} X(\omega)X^{-1}(\theta_i+)I_i].$$

Use the vector to obtain

$$x(t) = X(t)(\mathcal{I} - X(\omega))^{-1}[\int_0^\omega X(\omega)X^{-1}(s)f(s)ds + \sum_{0<\theta_i<\omega} X(\omega)X^{-1}(\theta_i+)I_i]$$
$$+ \int_0^t X(t)X^{-1}(s)f(s)ds + \sum_{0<\theta_i<t} X(t)X^{-1}(\theta_i+)I_i.$$

Now, use the periodicity of the solution and write

$$x(t) = X(t)[\int_0^\omega X(\omega)X^{-1}(s)f(s)ds + \sum_{0<\theta_i<\omega} X(\omega)X^{-1}(\theta_i+)I_i]+$$
$$\int_0^t X(t)X^{-1}(s)f(s)ds + \sum_{0<\theta_i<t} X(t)X^{-1}(\theta_i+)I_i =$$
$$\int_0^t [X(t)(\mathcal{I} - X(\omega))^{-1}X(\omega)X^{-1}(s) + X(t)X^{-1}(s)]f(s)ds+$$
$$\sum_{0<\theta_i<t} [X(t)(\mathcal{I} - X(\omega))^{-1}X(\omega)X^{-1}(\theta_i+) + X(t)X^{-1}(\theta_i+)]I_i+$$
$$\int_t^\omega X(t)(\mathcal{I} - X(\omega))^{-1}X^{-1}(s)f(s)ds + \sum_{t\le\theta_i<\omega} X(t)(\mathcal{I} - X(\omega))^{-1}X^{-1}(\theta_i+)I_i =$$
$$\int_0^t [X(t+\omega)(\mathcal{I} - X(\omega))^{-1}X^{-1}(s) + X(t)X^{-1}(s)]f(s)ds+$$

$$\sum_{0<\theta_i<t} [X(t+\omega)(\mathcal{I} - X(\omega))^{-1} X^{-1}(\theta_i+) + X(t)X^{-1}(\theta_i+)] I_i +$$

$$\int_t^{\omega} X(t)(\mathcal{I} - X(\omega))^{-1} X^{-1}(s) f(s) ds + \sum_{t\le\theta_i<\omega} X(t)(\mathcal{I} - X(\omega))^{-1} X^{-1}(\theta_i+) I_i =$$

$$\int_0^{\omega} G(t,s) f(s) ds + \sum_{0<\theta_i<\omega} G(t,\theta_i+) I_i .$$

The theorem is proved. □

Notes

Basics of linear impulsive systems and periodic equations were investigated in [78, 111, 138–142]. Linear nonhomogeneous systems with impulses, existence of periodic and almost periodic solutions using reduction to discrete equations and theory of generalized functions were investigated in [75]. Gram–Schmidt orthonormalization method for impulsive linear systems was applied in [33].

Chapter 5
Nonautonomous Systems with Variable Moments of Impulses

5.1 Description of Systems

Let $G \subset \mathbb{R}^n$ be an open and connected set, I an open interval in $\mathbb{R}$, and $\mathcal{A}$ an interval in $\mathbb{Z}$. We consider the following system:

$$\begin{aligned} x' &= f(t,x), \\ \Delta x|_{t=\tau_i(x)} &= J_i(x), \end{aligned} \tag{5.1}$$

where $(t,i,x) \in I \times \mathcal{A} \times G$, the function $f(t,x)$ is continuous on $I \times G$, functions J_i are defined on G, and $\tau_i(x), i \in \mathcal{A}$, are continuous on G functions.

The system combines the differential equation

$$x' = f(t,x), \tag{5.2}$$

and the equation of jumps

$$\Delta x|_{t=\tau_i(x)} = J_i(x). \tag{5.3}$$

The differential equation (5.2) satisfies the condition (M0), Chap. 2. Moreover, the following assumptions are fulfilled:

(N1) there exist positive numbers $\underline{\theta}, \bar{\theta}$ such that $\underline{\theta} < \tau_{i+1}(x) - \tau_i(x) < \bar{\theta}$ for all $i \in \mathcal{A}, x \in G$;
(N2) for all $i \in \mathcal{A}$ and $x \in G$, there exist real numbers $\alpha_i, \beta_i \in I$ such that $\alpha_i \leq \tau_i(x) \leq \beta_i$;
(N3) $\tau_i(x + J_i(x)) \leq \tau_i(x)$ for all $i \in \mathcal{A}$ and $x \in G$;
(N4) if $\xi(t, \tau_i(c), c + J_i(c)), c \in G, i \in \mathcal{A}$, is a solution of (5.2), then $t \neq \tau_i(\xi(t, \tau_i(c), c + J_i(c)))$ for all $t > \tau_i(c)$.
(N5) if $\eta(t, \tau_i(c), c), c \in G, i \in \mathcal{A}$, is a solution of (5.2), then $t \neq \tau_i(\eta(t, \tau_i(c), c))$ for all $t < \tau_i(c)$.

For simplicity of notation, denote by $\Gamma_i \subset I \times G$ the surface $t = \tau_i(x)$, and G_i subregion of $I \times G$ between Γ_i and $\Gamma_{i+1}, i \in \mathcal{A}$. More precisely, $G_i = \{(t,x) : \tau_i(x) < t \leq \tau_{i+1}(x), x \in G\}$.

M. Akhmet, *Principles of Discontinuous Dynamical Systems*,
DOI 10.1007/978-1-4419-6581-3_5,

The last five conditions are important for the next analysis. The condition (N1) guarantees that discontinuity moments of any solution of (5.1) format a B-sequence. Conditions (N1) with (N2) imply that each solution which intersects surfaces Γ_j and $\Gamma_k, j < k-1$, must intersect all surfaces $\Gamma_i, j < i < k$, between the two. By condition (N4), each solution of (5.1) may intersect every surface of discontinuity Γ_i at most once. Indeed, if θ_i is the first meeting moment of the solution $x(t)$ with the surface, then for $t > \theta_i$ the solution is equal to the solution $\xi(t, \theta_i, x(\theta_i) + J_i(x(\theta_i)))$ of (5.2), and it cannot intersect the surface again, since of (N4). The phenomenon, when each solution meets every surface of discontinuity at most once is called the "absence of beating", and we will discuss the concept in the next section. Conditions (N1),(N3) imply that the point $(t, x(t))$ after a meeting with a surface $t = \tau_i(x)$ jumps in the region between surfaces $t = \tau_i(x)$ and $t = \tau_{i+1}(x)$. In general, (N1)–(N4) produce circumstances which allow us to keep the order of intersection of the surfaces of discontinuity by a solution. This order helps us to have well-formulated assertions as well as comprehensive proofs of the theory. We shall need condition (N5) for the left extension of solutions. Condition (N1) can be weakened for some special cases. For example, condition $\inf_G \tau_i(x) \to \infty$, as $i \to \infty$, is sufficient if we consider the increasing t.

The equation of jumps (5.3) is different from that of Chap. 2 since the solutions have discontinuities not at moments of intersection with the planes $t = \theta_i$, but at moments of intersection with surfaces $t = \tau_i(x), i \in \mathcal{A}$. That is, the moments are not prescribed, and not known until one starts to look for a certain solution of the impulsive system. That is why, we shall call system (5.1) a system with variable moments of impulses. Obviously, different solutions of the system have, in general, different moments of discontinuities.

5.2 Existence, Uniqueness, and Extension

In this section, we consider extension of solutions of the impulsive system, local existence, and uniqueness theorems. It is useful if the reader remembers the definitions of maximal intervals and the extension of solutions made in the Sect. 2.1 as it helps to understand the next discussion better. To shorten it, in the sequel, we denote by $\phi(t, \kappa, z), (\kappa, z) \in I \times G$, a solution of ordinary differential equation (5.2) with $\phi(\kappa, \kappa, z) = z$, and let $x(t) = x(t, t_0, x_0)$ be a solution of (5.1) with $x(t_0) = x_0$. Suppose the point (t_0, x_0) does not belong to any of the surfaces of discontinuity. Let us say, it is lying between the surfaces Γ_j and Γ_{j+1}, for a fixed $j \in \mathcal{A}$ (see Fig. 5.1).

Similarly to the equation with fixed moments of impulses, we consider both directions of the extension:

(a) t is increasing and $t \geq t_0$;
(b) t is decreasing and $t \leq t_0$.

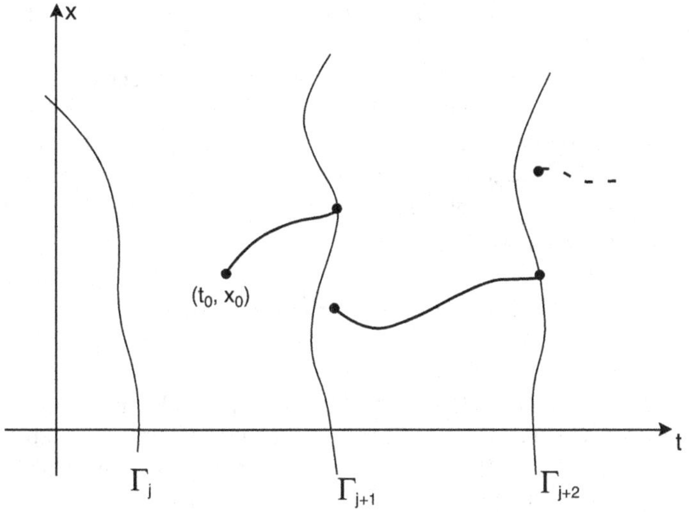

Fig. 5.1 Extension of a solution of system (5.1)

Consider the case (a). Let $[t_0, r), r \in \mathbb{R}$ or $r = \infty$, be the maximal right interval of existence of the solution $\phi(t, t_0, x_0)$. If $t \neq \tau_{j+1}(\phi(t, t_0, x_0))$ for all $t \in [t_0, r)$, then $[t_0, r)$ is the maximal interval of existence of $x(t)$. Otherwise, there exist solutions of equation $t = \tau_{j+1}(\phi(t, t_0, x_0))$. Denoting the least of them θ_{j+1}, we have that $x(t) = \phi(t, t_0, x_0)$ for $t \in [t_0, \theta_{j+1}]$. Particularly, $x(\theta_{j+1}) = \phi(\theta_{j+1}, t_0, x_0)$. Further, if $\Pi_{j+1}x(\theta_{j+1}) \in G$, then $x(\theta_{j+1}+) = \Pi_{j+1}x(\theta_{j+1})$. Taking into account that $(\theta_{j+1}, x(\theta_{j+1}+))$ is an interior point of $I \times G$, we can proceed the solution $\phi(t, \theta_{j+1}, x(\theta_{j+1}+))$ on some maximal interval $[\theta_{j+1}, r)$. Since of (N4), the last solution cannot meet the surface Γ_{j+1} again. There are two alternatives now. If $t \neq \tau_{j+2}(\phi(t, t_0, x_0)), t \in [t_0, r)$, then $[t_0, r)$ is the maximal right interval of existence of $x(t)$. Otherwise, we denote by θ_{j+2} the least solution of the equation $t = \tau_{j+2}(\phi(t, t_0, x_0))$, and $x(t) = \phi(t, \theta_{j+1}, x(\theta_{j+1}+))$ on $[\theta_{j+1}, \theta_{j+2}]$. Proceeding in this way one can find the right maximal interval of existence for $x(t)$.

Consider the case (b) now. That is, $t \leq t_0$. The point (t_0, x_0) lies, with a neighborhood, in G_j. By condition (M0), there is the left maximal interval of existence of $\phi(t, t_0, x_0)$. Denote by $(l, t_0]$ the interval. Consider the system

$$\begin{aligned} \Pi_j z &= \phi(t+, t_0, x_0), \\ t &= \tau_j(z), \end{aligned} \tag{5.4}$$

where $\phi(t+, t_0, x_0)$ is the right limit at the moment t, if it exists. If (5.4) is not solvable with respect to $t \in (l, t_0]$ and $z \in G$, and $t \neq \tau_j(\phi(t, t_0, x_0))$ for all t, $l < t \leq t_0$, then $(l, t_0]$ is a left maximal interval of $x(t)$. Otherwise, we denote by s the maximal among numbers t such that $t = \tau_j(\phi(t, t_0, x_0))$ and $t \in (l, t_0]$. Now, if

(5.4) is not solvable with respect to $t \in [s, t_0]$ and $z \in G$, then $(s, t_0]$ is a left maximal interval of $x(t)$. Otherwise, if (5.4) has a solution, (it may have several solutions, and even infinitely many solutions), then choose one of them and denote it by $\theta_j, x(\theta_j)$. We set $x(t) = \phi(t, t_0, x_0)$ for all $t \in (\theta_j, t_0]$. Particularly, $x(\theta_j+) = \phi(\theta_j, t_0, x_0)$. Next, consider the solution $\phi(t, \theta_j, x(\theta_j))$ to the left of $t = \theta_j$. This solution will not meet the surface Γ_j again. (Why?) Let $(l, \theta_j], l \in \mathbb{R}$ or $l = \infty$, be the left maximal interval of existence of $\phi(t, \theta_j, x(\theta_j))$. If $t \neq \tau_{j-1}(\phi(t, \theta_j, x(\theta_j))$ for all $t, l < t \leq \theta_j$, and the equation

$$\begin{aligned} \Pi_{j-1} z &= \phi(t+, \theta_j, x(\theta_j)), \\ t &= \tau_{j-1}(z), \end{aligned} \tag{5.5}$$

does not have a solution $t \in (l, \theta_j], z \in G$, then $(l, t_0]$ is a left maximal interval of existence of $x(t)$. Otherwise, we denote by s the maximal among numbers t such that $t = \tau_{j-1}(\phi(t, \theta_j, x(\theta_j)))$ and $t \in (l, \theta_j]$. If (5.5) is not solvable with respect to $t, z, t \in [s, t_0]$, then a left maximal interval of existence of $x(t)$ is $(s, t_0]$. Otherwise, (5.5) has solutions, and we choose one of them and denote it by $\theta_{j-1}, x(\theta_{j-1})$. Moreover, $x(t) = \phi(t, \theta_j, x(\theta_j)), t \in (\theta_{j-1}, \theta_j]$. Proceeding the discussion in this way, we shall determine a left maximal interval of $x(t)$.

Unite the left and right maximal intervals to obtain the maximal interval of $x(t)$.

Example 5.2.1. Discuss the problem of a maximal interval of $x(t) = x(t, t_0, x_0)$ if (t_0, x_0) belongs to one of the surfaces Γ_i.

Solution. Fix i such that $t_0 = \tau_i(x_0)$. If t is increasing, and $x_0 + J_i(x_0) \in G$, then the solution starts with the jump $x(t_0+) = x_0 + J_i(x_0)$. So, the solution is continuable to the right, as it was described above, in the case (a). Otherwise, $x_0 + J_i(x_0) \notin G$, and the right end-point of the maximal interval of existence is t_0. If t decreases, then conditions (N1),(N2) imply that the solution exists for $t < t_0$, and the further discussion is the same as that of case (b).

From the last discussion it follows that each solution $x(t)$ of (5.1) is a function from $\mathcal{PC}(J, \theta)$, where J is the interval of existence of this solution, and $\theta \subset J$ is the sequence of discontinuity moments of the solution.

Remark 5.2.1. The moment of intersection of a solution $x(t)$ and a surface Γ_i is a moment of discontinuity. Even if $J_i(x(\theta_i)) = 0$.

Next three assertions are generalized versions of Theorems 2.3.2, 2.3.3, and 2.3.4, and they can be proved easily if one yields the analysis just made above.

Theorem 5.2.1. *(Local existence theorem) Suppose the function $f(t, x)$ is continuous on $I \times G$, and $\Pi_i G \subseteq G, i \in \mathcal{A}$. Then for each $(t_0, x_0) \in I \times G$, there is a number $\alpha > 0$, such that a solution $x(t, t_0, x_0)$ of (5.1) exists in the open interval $(t_0 - \alpha, t_0 + \alpha)$.*

Theorem 5.2.2. *Suppose conditions (M0),(N1)–(N5) are valid. Then each solution of (5.1) has a maximal interval of existence and for this interval the following alternatives are possible:*

(1) it is an open set (α, β) such that any limit point of the set $(t, x(t))$ as t tends to an end point of the maximal interval is a boundary point of $I \times G$;
(2) it is a half-open interval $(\alpha, \beta]$, where $\beta = \tau_j(x(\beta))$ for some $j \in \mathcal{A}$, and any limit point of the set $(t, x(t))$ as $t \to \alpha$ is a boundary point of $I \times G$;
(3) it is a half-open interval $(\alpha, \beta]$, where $\alpha = \tau_i(x(\alpha))$ and $\beta = \tau_j(x(\beta))$ for some $i, j \in \mathcal{A}$. The limit $x(\alpha+)$ exists and it is an interior point of G.
(4) it is an open set (α, β) such that any limit point of the set $(t, x(t))$ as $t \to \beta$ belongs to the boundary of $I \times G, \alpha = \tau_i(x(\alpha))$, the limit $x(\alpha+)$ exists and it is an interior point of G.

In the sequel, the following condition plays an important role.

(N6) If $x(t)$ is a solution of (5.1) and $\theta_j, j \in \mathcal{A}$, is a discontinuity moment, then the system

$$\Pi_{j-1}z = \phi(t+, \theta_j, x(\theta_j)),$$
$$t = \tau_{j-1}(z), \tag{5.6}$$

has at most one solution $(t, z), t < \theta_j, z \in G$. If θ_k is the maximal discontinuity moment of $x(t)$, then the system

$$\Pi_k z = \phi(t+, s, x(s)),$$
$$t = \tau_k(z), \tag{5.7}$$

where $s > \theta_k$, has at most one solution $(t, z), t < s, z \in G$.

Theorem 5.2.3. *(Uniqueness theorem) Suppose $f(t, x)$ satisfies a local Lipschitz condition, and $x(t)$ is a solution of (5.1), which satisfies condition (N6). If $y(t)$ is a solution of (5.1) such that $y(\tau) = x(\tau)$ for some $\tau \in I$, then $x(t) = y(t)$ for all t, where both $x(t)$ and $y(t)$ are defined.*

Exercise 5.2.1. Prove the last three theorems.

Summarizing, one can conclude that the following assertion is valid.

Theorem 5.2.4. *Assume that conditions (M0),(N1)–(N5) are fulfilled. Then every solution $x(t) = x(t, t_0, x_0), (t_0, x_0) \in I \times G$, of (5.1) has a maximal interval of existence. The solution is unique if condition (N6) is valid. The interval is left-open and either right-open or right-closed.*

5.3 Beating Phenomena and Related Properties

The task of investigation of the global behavior of solutions for equations with non-fixed moments of impulses is more complex than that of systems with impulses acting at prescribed moments. A reason for this is the possibility of the 'beating' of solutions against the surfaces of discontinuity.

Let $x(t)$ be a solution of (5.1). Fix an integer j and assume that $x(t)$ meets the surface Γ_j more than once. In other words, assume that equation $t = \tau_j(x(t))$ has several solutions with respect to t. They are not equal to each other. It is obvious that if we try to find an integral equation equivalent to (5.1), then the expression $J_j(x(t))$, where j is fixed, will be involved in that equation more than one time with various values of t. This will create certain difficulties in the analysis, exceptionally if the number of the members is not predictable. The number is important if we analyze stability problems, periodicity of solutions, almost periodicity, etc. That is why it is necessary to find conditions which will provide possibility for solutions to meet the surfaces of discontinuity not more than once, or, more precisely, exactly once. Conditions (N1),(N3) imply that the point $(t, x(t))$ after the first meeting with a surface $t = \tau_i(x)$ jumps inside the region between surfaces $t = \tau_i(x)$ and $t = \tau_{i+1}(x)$. By (N4), the integral curve does not intersect the surface Γ_i again. Obviously, there should be easily verifiable conditions, which guarantee (N4). There exist several results on the subject of absence of beating [63, 142]. We are going to consider simple and effective one, which is useful in applications, and which enables to understand the essence of the phenomena easily.

Let us formulate the following additional conditions:

(N7)
$$\|f(t,x) - f(t,y)\| \le l_f \|x - y\| \quad for\,all \quad x, y \in G;$$

(N8)
$$\|J_i(x) - J_i(y)\| \le l_J \|x - y\| \quad for\,all \quad i \in \mathcal{A}, x, y \in G;$$

(N9)
$$|\tau_i(x) - \tau_i(y)| \le l_\tau \|x - y\| \quad for\,all \quad i \in \mathcal{A}, x, y \in G.$$

In (N7)–(N9), l_f, l_J, l_τ are positive constants.

(N10) The inequality
$$l_\tau M_f < 1$$
is valid, where $M_f = \sup\limits_{(t,x)\in I\times G} \|f(t,x)\| < \infty.$

Lemma 5.3.1. *Suppose the surfaces $\Gamma_i, i \in \mathcal{A}$, and the function $f(t,x)$ satisfy conditions (N9) and (N10). Then conditions (N4) and (N5) are valid.*

Proof. Let us prove that $(N4)$ is valid. Assume on the contrary that there is a solution $\xi(t) = \xi(t, s, c + J_j(c)), c \in G, s = \tau_j(c), j \in \mathcal{A}$, of (5.2), which intersects surface Γ_j at a moment $s_1, s < s_1$. We have that

$$0 < s_1 - s = \tau_j(x(s_1)) - \tau_j(x(s)) \le$$

$$l_\tau \| \int_s^{s_1} f(t, \xi(t)) dt \| \le M_f l_\tau (s_1 - s).$$

The last inequality contradicts (N9).

Exercise 5.3.1. Prove that (N5) is valid, similarly to that of (N4).

The lemma is proved. □

Lemma 5.3.2. *Assume that conditions (N1) and (N2) are fulfilled and* $x(t) : I \to G$ *is a solution of (5.1). Then* $x(t)$ *intersects every surface* $\Gamma_i, i \in \mathcal{A}$.

Proof. Assume on the contrary that $x(t)$ does not intersect Γ_j for some $j \in \mathcal{A}$. Condition (N1) implies, without loss of generality, that we may assume the surface as a unique surface of discontinuity. In other words, $\mathcal{A} = \{j\}$. Introduce a new function $r(t) = t - \tau_j(x(t))$. As (N2) is valid, there exist $\alpha, \beta, \alpha < \alpha_j \le \beta_j < \beta$, such that $r(\alpha) < 0 < r(\beta)$, and by the continuity of $r(t)$ there exists a point $\zeta \in (\alpha, \beta)$ such that $r(\zeta) = 0$. That is, $\zeta = \tau_j(x(\zeta))$. The lemma is proved. □

Using the last two lemmas one can formulate the following assertion.

Theorem 5.3.1. *Assume that conditions (N1)–(N3), (N8), and (N9) are fulfilled. Then every solution* $x(t) : I \to G$ *of (5.1) intersects each of the surfaces* $\Gamma_i, i \in \mathcal{A}$ *exactly once.*

Besides the last theorem the following general assertion can be formulated.

Theorem 5.3.2. *Assume that conditions (N1)–(N4) are fulfilled. Then every solution* $x(t) : I \to G$ *of (5.1) intersects each of the surfaces* $\Gamma_i, i \in \mathcal{A}$, *exactly once.*

Remark 5.3.1. In this book we discuss only systems, which satisfy conditions of absence of the beating, that is, each solution of a system intersects every surface of discontinuity not more than once.

Exercise 5.3.2. Solve the following problems.

1. Consider the impulsive system

$$\begin{aligned} x' &= 0, \\ \Delta x|_{t=i+l|x|} &= -\frac{1}{2}x, \end{aligned} \tag{5.8}$$

   where $t, x \in \mathbb{R}, i \in \mathbb{Z}, l$ is a fixed positive number. Prove that there is no beating of solutions against the surfaces of discontinuity.
2. Let a solution $\phi(t) : I \to G, \phi(t_0) = x_0$, of (5.1) intersects each of the surfaces $\Gamma_i, i \in \mathcal{A}$, exactly once at the moment $t = \theta_i$. Prove that

$$\phi(t) = \begin{cases} x_0 + \int\limits_{t_0}^{t} f(s,\phi(s))ds + \sum\limits_{t_0 \le \theta_i < t} J_i(\phi(\theta_i)), & t \ge t_0, \\ x_0 + \int\limits_{t_0}^{t} f(s,\phi(s))ds - \sum\limits_{t \le \theta_i < t_0} J_i(\phi(\theta_i)), & t < t_0. \end{cases} \tag{5.9}$$

3. Assume that conditions of Theorem 5.3.1 are valid and a function $\phi(t) : I \to G$ satisfies (5.9), where θ_i are all points of I such that $\theta_i = \tau_i(\phi(\theta_i))$. Prove that $\phi(t)$ is a solution of (5.1) on I.

On the basis of results of the last two exercises, one can formulate the following assertion.

Theorem 5.3.3. *Assume that conditions (N1)–(N5) are fulfilled. A function* $x(t) : I \to G, x(t_0) = x_0, (t_0, x_0) \in I \times G$, *with the sequence of discontinuity moments* $\{\theta_i\} \subset I, i \in \mathcal{A}$, *is a solution of (5.1) on* I *if and only if it satisfies the following integral equation:*

$$x(t) = \begin{cases} x_0 + \int\limits_{t_0}^{t} f(s, x(s))ds + \sum\limits_{t_0 \le \theta_i < t} J_i(x(\theta_i)), & t \ge t_0, \\ x_0 + \int\limits_{t_0}^{t} f(s, x(s))ds - \sum\limits_{t \le \theta_i < t_0} J_i(x(\theta_i)), & t < t_0. \end{cases} \tag{5.10}$$

Remark 5.3.2. Obviously, that investigation of the system (5.1) through integral equations (5.10) is a complex work as the moments of discontinuity are not prescribed. In the next sections, we provide a method which will help to investigate (5.1) on the basis of an integral equation.

Theorem 5.3.4. *Assume that conditions (N8),(N9) are fulfilled and*

$$l_J < 1, \qquad M_f l_\tau < 1 - l_J. \tag{5.11}$$

Then (N6) is valid.

Proof. Assume on the contrary that (N6) is not valid for a solution $x(t)$ and some $j \in \mathcal{A}$. Then there exist vectors $z_1, z_2, z_1 \neq z_2$, and moments $t_1, t_2, t_2 \ge t_1$, such that $z_1 + J_j(z_1) = x(t_1), t_1 = \tau_j(z_1)$, and $z_2 + J_j(z_2) = x(t_2), t_2 = \tau_j(z_2)$. If $t_2 > t_1$, then

$$\begin{aligned} \|z_1 - z_2\|(1 - l_J) \le \|z_1 - z_2 + J_j(z_1) - J_j(z_2)\| &= \| \textstyle\int_{\tau_j(z_1)}^{\tau_j(z_2)} f(s, x(s))ds\| \\ &\le l_\tau M_f \|z_1 - z_2\|. \end{aligned}$$

The last inequality contradicts (5.11). If $t_2 = t_1$, then

$$\|z_1 - z_2\|(1 - l_J) \le \|z_1 - z_2 + J_j(z_1) - J_j(z_2)\| = \|x(t_1) - x(t_2)\| = 0.$$

We have another contradiction. The theorem is proved. □

5.4 The Topology on the Set of Discontinuous Functions

A difficulty of investigation of system (5.1) is that the moments of discontinuity of distinct solutions do not, in general, coincide. To investigate neighborhoods of solutions of differential equations with impulses at variable moments, we introduce the following concepts of closeness for piecewise continuous functions. Various metrics and topologies for discontinuous functions are described in [7, 32, 75, 90, 142, 147].

Denote by $\widehat{[a,b]}, a,b \in \mathbb{R}$, the interval $[a,b]$, whenever $a \leq b$ and $[b,a]$, otherwise.

Definition 5.4.1. Solutions $x_1(t) \in \mathcal{PC}(T_1, \theta^1)$ and $x_2(t) \in \mathcal{PC}(T_2, \theta^2)$ are said to be ϵ-equivalent if:

(1) the measure of the symmetric difference of the domains T_1 and T_2 does not exceed ϵ;
(2) $|\theta_i^1 - \theta_i^2| < \epsilon$ for all i;
(3) the inequality $\|x_1(t) - x_2(t)\| < \epsilon$ is valid for all t, which satisfy $t \notin \widehat{[\theta_i^1, \theta_i^2]}$ for all i.

Definition 5.4.2. A solution $x_1(t)$ is in the ϵ-neighborhood, $\epsilon > 0$, of a solution $x_2(t) \in \mathcal{PC}(T_2, \theta^2)$, if either it is continuable on an interval T_1, or can be restricted to an interval, such that the two are ϵ-equivalent.

If $x_1(t)$ and $x_2(t)$ are ϵ-equivalent then $x_1(t)$ is in the ϵ-neighborhood of $x_2(t)$, and vice versa.

The equivalence of two piecewise continuous functions, when ϵ is small, means roughly that they have close discontinuity points, and the values of the functions are close at points that do not lie on intervals between the corresponding discontinuity points of these functions.

The topology defined with the aid of ϵ-neighborhoods is called the B-topology. One can easily see that it is Hausdorff and it can be considered also if two solutions x_1 and x_2 are defined on a semi-axis or on the entire real axis.

5.5 B-Equivalence: General Case

Consider (5.1), assuming that conditions (N1)–(N3), (N7)–(N10), are valid, $\tau_i(x) = \theta_i + \kappa_i(x)$, with $|\kappa_i(x)| < \nu$, for some positive number ν and for all $x \in G, i \in \mathcal{A}$. Fix a number i. Let $x_0(t), x_0(\theta_i) = x$, be a solution of the system

$$x' = f(t,x). \tag{5.12}$$

Denote by ξ_i the meeting moment of the solution with the surface of discontinuity so that $\xi_i = \theta_i + \kappa_i(x_0(\xi_i))$. Let also, $x_1(t)$ be a solution of (5.12) such that $x_1(\xi_i) = x_0(\xi_i) + J_i(x_0(\xi_i))$. Assume that it exists on the interval $\widehat{[\theta_i, \xi_i]}$ and define the following map (see Fig. 5.2),

$$W_i(x) = \int_{\theta_i}^{\xi_i} f(u, x_0(u))du + J_i(x + \int_{\theta_i}^{\xi_i} f(u, x_0(u))du) + \int_{\xi_i}^{\theta_i} f(u, x_1(u))du. \tag{5.13}$$

The map W_i defined for each i is called the B-map.

Fig. 5.2 Construction of the map W_i

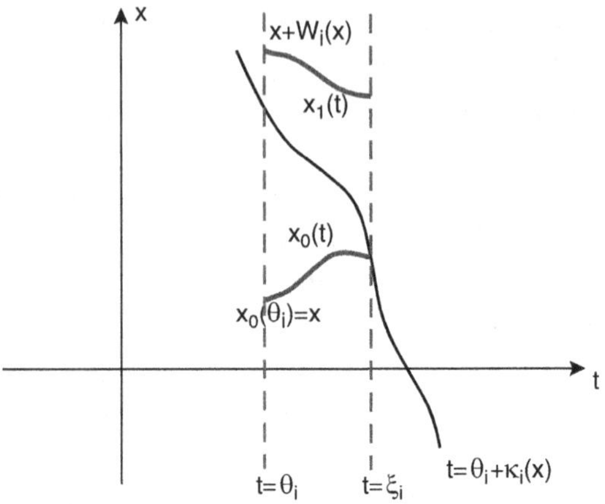

Beside (5.1) consider the following system:

$$\begin{aligned} x' &= f(t, x), \\ \Delta x|_{t=\theta_i} &= W_i(x). \end{aligned} \tag{5.14}$$

We say that systems (5.1) and (5.14) are *B-equivalent* in $G \subset \mathbb{R}^n$ if there exists a set $G_1 \subset G$ such that for each solution $x(t)$ of (5.1) defined on an interval U, with discontinuity moments ξ_i and $x(t) \in G_1, t \in U$, there exists a solution $y(t) : U \to G$ of (5.14), satisfying

$$x(t) = y(t), t \notin (\widehat{\xi_i, \theta_i}] \tag{5.15}$$

if $\xi_i \neq \theta_i$, and

$$x(\xi_i) = y(\theta_i) \tag{5.16}$$

if $\xi_i = \theta_i$, for all i. In particular,

$$x(\theta_i) = y(\theta_i+), x(\xi_i) = y(\xi_i) \text{ if } \theta_i > \xi_i, \tag{5.17}$$

$$x(\theta_i) = y(\theta_i), x(\xi_i+) = y(\xi_i) \text{ if } \theta_i < \xi_i. \tag{5.18}$$

Conversely, for each solution $y(t)$ of (5.14) with $y(t) \in G_1, t \in U$, there exists a solution $x(t), t \in U$, of (5.1) such that (5.15)–(5.18) are valid.

Remark 5.5.1. To define W_i it is sufficient if the solution $x_1(t)$ is defined on the left-open interval $(\widehat{\theta_i, \xi_i}]$.

Suppose

$$\sup_{I \times G} \|f(t, x)\| + \sup_{\mathcal{A} \times G} \|J_i(x)\| = M < \infty. \tag{5.19}$$

Fix a vector $x_0 \in G$, and real positive numbers $h, H = h + \nu M, \bar{H} = h + (1 + 2\nu)M$. We assume that the set $\|x - x_0\| \leq \bar{H}$ is included in G.

Theorem 5.5.1. *Suppose (5.1) satisfies all conditions listed above. Then:*

(1) functions $W_i, i \in \mathcal{A}$, are defined on the set $G_h = \{x \in \mathbb{R}^n : \|x - x_0\| \leq h\}$;
(2) for arbitrary $x, y \in G_h$ and every $i \in \mathcal{A}$ it is true that

$$\|W_i(x) - W_i(y)\| \leq k(l_f, l_J, l_\tau)\|x - y\|, \tag{5.20}$$

where

$$k(l_f, l_J, l_\tau) = l_f \nu[(1 + l_J)e^{\nu l_f} + e^{2\nu l_f}(1 + l_J(1 - M l_\tau))(1 - M l_\tau)^{-1}] +$$

$$l_J + l_\tau(2 + l_J)e^{\nu l_f} M(1 - M l_\tau)^{-1};$$

(3) systems (5.1) and (5.14) are B-equivalent in $G_H = \{x \in \mathbb{R}^n : \|x - x_0\| \leq H\}$, such that if a solution of (5.1) or (5.14) get values in G_h, then the corresponding solution get values in G_H.

Proof. Fix i, and let $x_0(t)$ and $x_1(t)$ be solutions of (5.12), which are mentioned in (5.13). We have

$$x_0(t) = x + \int_{\theta_i}^{t} f(u, x_0(u))du, \quad t \in \widehat{[\theta_i, \xi_i]}.$$

Assuming $x \in G_h$, and then using the last equation one can find that $\|x_0(t) - x_0\| < h + \nu M = H$, if $t \in \widehat{[\theta_i, \xi_i]}$. Similarly, one check that $\|x_1(t) - x_0\| < \bar{H}$. Thus, W_i is defined on G_h and part (1) of the theorem is proved. □

We fix $y \in G_h$, and next, determine $W_i(y)$. Denote by $y_0(t), y_0(\theta_i) = y$, the solution of (5.12), and η_i the meeting moment of $y_0(t)$ with Γ_i. Let $y_1(t)$ be the solution of (5.12), which satisfies $y_1(\eta_i) = y_0(\eta_i) + J_i(y_0(\eta_i)), t \in [\theta_i, \eta_i]$. Then

$$W_i(y) = \int_{\theta_i}^{\eta_i} f(u, y_0(u))du + J_i(y + \int_{\theta_i}^{\eta_i} f(u, y_0(u))du) +$$

$$\int_{\eta_i}^{\theta_i} f(u, y_1(u))du. \tag{5.21}$$

Without loss of generality assume that $\theta_i \leq \xi_i \leq \eta_i$. Using the Gronwall–Bellman Lemma, one can find that

$$\|x_0(t) - y_0(t)\| \leq e^{\nu l_f}\|x - y\|, \tag{5.22}$$

if $t \in [\theta_i, \xi_i]$. Then, applying (5.22) and condition (N10) we obtain

$$\eta_i - \xi_i \leq l_\tau e^{\nu l_f}(1 - M l_\tau)^{-1}\|x - y\|. \tag{5.23}$$

Moreover, using definition of solutions x_1 and y_1, we get that

$$\|x_1(t) - y_1(t)\| \le e^{2\nu l_f}(1 + l_J(1 - Ml_\tau))(1 - Ml_\tau)^{-1}\|x - y\|, \tag{5.24}$$

if $t \in [\theta_i, \xi_i]$.

Now, subtracting (5.21) from (5.13) yields

$$W_i(x) - W_i(y) = \int_{\theta_i}^{\xi_i} f(u, x_0(u))du - \int_{\theta_i}^{\eta_i} f(u, y_0(u))du +$$

$$J_i(x + \int_{\theta_i}^{\xi_i} f(u, x_0(u))du) - J_i(y + \int_{\theta_i}^{\eta_i} f(u, y_0(u))du) +$$

$$\int_{\xi_i}^{\theta_i} f(u, x_1(u))du - \int_{\eta_i}^{\theta_i} f(u, y_1(u))du =$$

$$\int_{\theta_i}^{\xi_i} [f(u, x_0(u)) - f(u, y_0(u))]du + \int_{\xi_i}^{\eta_i} f(u, y_0(u))du +$$

$$J_i(x + \int_{\theta_i}^{\xi_i} f(u, x_0(u))du) - J_i(y + \int_{\theta_i}^{\eta_i} f(u, y_0(u))du) +$$

$$\int_{\theta_i}^{\xi_i} [f(u, x_1(u)) - f(u, y_1(u))]du + \int_{\xi_i}^{\eta_i} f(u, y_1(u))du,$$

and, using (5.22)–(5.24), we have that

$$\|W_i(x) - W_i(y)\| \le \int_{\theta_i}^{\xi_i} \|f(u, x_0(u)) - f(u, y_0(u))\|du + \int_{\xi_i}^{\eta_i} \|f(u, y_0(u))\|du +$$

$$l_J[\|x - y\| + \int_{\theta_i}^{\xi_i} \|f(u, x_0(u)) - f(u, y_0(u))\|du + \int_{\xi_i}^{\eta_i} \|f(u, y_0(u))\|du] +$$

$$\int_{\theta_i}^{\xi_i} \|f(u, x_1(u)) - f(u, y_1(u))\|du + \int_{\xi_i}^{\eta_i} \|f(u, y_1(u))\|du \le$$

$$\int_{\theta_i}^{\xi_i} l_f\|x_0(u) - y_0(u)\|du + M(\eta_i - \xi_i) + l_J[\|x - y\|$$

$$+ \int_{\theta_i}^{\xi_i} l_f\|x_0(u) - y_0(u)\|du + M(\eta_i - \xi_i)] +$$

$$\int_{\theta_i}^{\xi_i} l_f\|x_1(u) - y_1(u)\|du + M(\eta_i - \xi_i) \le k(l_f, l_J, l_\tau)\|x - y\|.$$

That is, (5.20) is valid. B-equivalence of the systems follows immediately from the definition of W_i.

Now, let us prove that $x(t) \in G_h, t \in I$, implies $y(t) \in G_H, t \in I$. Indeed, for all $t \notin [\widehat{\xi_i, \eta_i}]$, we have $y(t) = x(t)$. Hence, it is sufficient to consider an interval $[\widehat{\xi_i, \eta_i}]$, for some fixed i. Assume, for the sake of simplicity, that $\xi_i < \eta_i$. Then $y(\xi_i) = x(\xi_i)$ and $y(t) = x(\xi_i) + \int_{\xi_i}^{t} f(u, y(u))du, t \in (\xi_i, \eta_i]$. Finally,

$$\|y(t)\| \leq \|x(\xi_i)\| + \int_{\xi_i}^{t} M du \leq h + M\nu = H,$$

if $t \in (\xi_i, \eta_i]$.

Similarly one can check that $y(t) \in G_h, t \in I$, implies $x(t) \in G_H$, for all $t \in I$. The theorem is proved. □

Exercise 5.5.1. Interpret results of the last theorem with $\nu = 0$ and $l_\tau = 0$: evaluate $k(l_f, l_J, l_\tau)$ and W_i.

Example 5.5.1. Determine a B-equivalent system for the following impulsive equation:

$$\begin{aligned} x' &= 2x, \\ \Delta x|_{t \neq i+x} &= -\sin^2(x), \end{aligned} \tag{5.25}$$

where $t, x \in \mathbb{R}, i \in \mathbb{Z}$.

Solution. Fix $i \in \mathbb{Z}$. We have that $x_0(t) = x(t, i, x) = x\mathrm{e}^{2(t-i)}$ is a solution of

$$x' = 2x. \tag{5.26}$$

The moment of intersection of this solution with the line $t = i + x$ has to be found from $t = i + x\mathrm{e}^{2(t-i)}$. It is equal to $\zeta_i = i + \phi(x)$, where $u = \phi(x)$ is an implicit function such that

$$u = x\mathrm{e}^{2u}. \tag{5.27}$$

One can easily see that ϕ is a continuously differentiable function. Thus, we have that $x_0(\zeta_i) = x\mathrm{e}^{2\phi(x)}$. Similarly, one can evaluate the solution $x_1(t) = x(t, \zeta_i, x_0(\zeta_i) - \sin^2(x_0(\zeta_i)) = (x_0(\zeta_i) - \sin^2(x_0(\zeta_i)))\mathrm{e}^{2(t-\zeta_i)}$ of (5.26). Finally, one can find that $W_i(x) = x_1(i) - x = -\sin^2(x\mathrm{e}^{2\phi(x)}))\mathrm{e}^{-2\phi(x)}$, and the equivalent system is

$$\begin{aligned} y' &= 2y, \\ \Delta y|_{t \neq i} &= -\sin^2(y\mathrm{e}^{2\phi(y)})\mathrm{e}^{-2\phi(y)}. \end{aligned} \tag{5.28}$$

Remark 5.5.2. The method of investigation of qualitative properties of a differential equation by a reduction to a more simple system is very fruitful for any kind of equations. One can mention, for instance, the Floquet–Lyapunov transformation of linear periodic systems, or the method of the first approximation, which not necessarily must be the approximation by a linear system. Generally speaking, the B-equivalence considered in our book is a first approximation. To describe our approach better, let us start with a more general case. Consider beside system (5.1), another one,

$$\begin{aligned} y' &= g(t, y), \\ \Delta y|_{t \neq \upsilon_i(y)} &= W_i(y). \end{aligned} \tag{5.29}$$

Assuming that system (5.29) has a specified quality, and differences $f - g$, $\upsilon - \tau$, $W - J$, are small in some sense, one may make conclusion about that property for system (5.1). One can call (5.29) a *first approximation system.* If we restrict ourself on asymptotic features of the equation, then one can say about *comparison method* for systems (5.1) and (5.29) as usually it is named for ordinary differential equations. We shall use the first approximation systems not only to investigate asymptotic properties but also for periodic and bounded solutions, smoothness of solutions, and even chaotic behavior of discontinuous dynamics. Since the equations with fixed moments of impulses have been well-investigated, it is beneficial to use them for the approximation. Moreover, we suppose that it will be possible to consider approximations of higher order in future. So, it is natural to call the approach the method of reduction to systems with fixed moments of impulses.

Assume that $\upsilon_i(y) \equiv \theta_i$ in (5.29), where θ_i is a sequence of ordered real numbers, and consider the following equation:

$$\begin{aligned} y' &= g(t, y), \\ \Delta y|_{t \neq \theta_i} &= W_i(y). \end{aligned} \tag{5.30}$$

Let us show, how one may choose the sequence θ. If (5.1) has the zero solution, then $\theta_i = \tau_i(0)$. If one investigates stability of a solution $\phi(t)$ of (5.1), then it is convenient to set the moments of discontinuity of this solution as θ_i, and rewrite the surfaces of discontinuity as $\tau_i(x) = \theta_i + \kappa_i(x)$, where $\kappa_i(x) = \tau_i(x) - \theta_i$.

5.6 Continuity Properties

In this section, we discuss the continuous dependence by using the B-equivalence method and the topology in the set of piecewise continuous functions.

Consider system (5.1) on the set $\Omega = I \times \mathcal{A} \times G$ of points (t, i, x), where $G \subset \mathbb{R}^n$ is an open and connected set, I is an open interval in $\mathbb{R}$, and $\mathcal{A}$ is an interval in $\mathbb{Z}$. Denote by $\tilde{I} = [a, b] \subset I$ a closed finite interval and fix it. Let $x(t)$ be a solution of (5.1) such that $x(t) : \tilde{I} \to G, x(t_0) = x_0, (t_0, x_0) \in \tilde{I} \times G$, and let $\theta_i, i = m, \ldots, k$, be moments of discontinuity of the solution, that is, $\theta_i = \tau_i(x(\theta_i)), i = m, \ldots, k$. Assume that for all $x \in G, a < \tau_m(x) < \ldots < \tau_k(x) < b$, and denote $G_i = \{(t, x) : \tau_i(x) < t \leq \tau_{i+1}(x), x \in G\}, i = m, \ldots, k - 1, G_{m-1} = \{(t, x) : a < t \leq \tau_m(x), x \in G\}, G_k = \{(t, x) : \tau_k(x) < t \leq b, x \in G\}$. The conditions imply that there are small neighborhoods of points $(\theta_i, x(\theta_i))$, where the surfaces of discontinuity have the form $\tau_i(x) = \theta_i + \kappa_i(x)$, where $\kappa_i(x) = \tau_i(x) - \theta_i$ and $\kappa_i(x(\theta_i)) = 0$. There exists a positive number ν such that $|\kappa_i(x)| < \nu$ for all i, and the number ν is arbitrary small if the neighborhoods are chosen sufficiently small. Thus, we can claim that all conditions of the last section needed for the reduction to the system with fixed moments of impulses, that is, $\theta_i, i \in \mathcal{A}$, are valid.

Let $D(t_0, \delta) = \{(t_0, x) : \|x - x_0\| < \delta\}$ be a disc centered at (t_0, x_0), and $d(x_0, \delta) = \{(t, x_0) : t_0 - \delta < t < t_0 + \delta\}$ be an interval centered at (t_0, x_0) with the radius $\delta > 0$. Fix a number j such that $(t_0, x_0) \in G_j$. We shall need the following definitions.

Definition 5.6.1. The solution $x(t)$ of (5.1) continuously depends on x_0 if for an arbitrary $\epsilon > 0$ there exists $\delta > 0$ such that any other solution $\tilde{x}(t)$ of (5.1) is in the ϵ-neighborhood of $x(t)$, as soon as $(t_0, \tilde{x}(t_0)) \in D(t_0, \delta) \cap G_j$.

Definition 5.6.2. The solution $x(t)$ of (5.1) continuously depends on t_0 if to any $\epsilon > 0$ there corresponds $\delta > 0$ such that any other solution $\tilde{x}(t), \tilde{x}(\tilde{t_0}) = x_0$, of (5.1) is in the ϵ-neighborhood of $x(t)$, as soon as $(\tilde{t_0}, x_0) \in d(x_0, \delta) \cap G_j$.

Using Theorems 2.7.1, 2.7.2, and reduction to the equation with fixed moments of impulses, one can easily prove that the following assertions, Theorems 5.6.1–5.6.4, are valid.

Theorem 5.6.1. *Suppose conditions (N1)–(N3), (N7)–(N10), are fulfilled. Then the solution $x(t)$ of (5.1) continuously depends on t_0 and x_0, if $t_0 = a$.*

Theorem 5.6.2. *Suppose conditions (N1)–(N3), (N7)–(N10), (5.11) are fulfilled, and $k(l_f, l_J, l_\tau) < 1$. Then each solution $x(t) = x(t, t_0, x_0), (t_0, x_0) \in I \times G$, of (5.1) is continuous in t_0 and x_0 on each closed finite interval, where it is defined.*

Remark 5.6.1. Conditions $(t_0, \tilde{x}_0) \in D(t_0, \delta) \cap G_j$ and $(\tilde{t_0}, x_0) \in d(x_0, \delta) \cap G_j$ in Definitions 5.6.1 and 5.6.2 are important. That is, assumptions $(t_0, \tilde{x}_0) \in D(t_0, \delta)$ or $(\tilde{t_0}, x_0) \in d(x_0, \delta)$ are not sufficient for the constructive description of the dependence. Indeed, consider the following simple system,

$$\begin{aligned} x' &= 0, \\ \Delta x|_{t = x^{1/2}} &= -5, \end{aligned} \tag{5.31}$$

where $x, t \in \mathbb{R}$. Fix solutions $x(t) = x(t, 2, 4)$ and $x_1(t) = x(t, 2, 4 - \kappa), 0 < \kappa < 4$. One can easily see that the distance between $x(t)$ and $x_1(t)$ near $t = 2$ tends to 5 as $\kappa \to 0$.

Exercise 5.6.1. Show that the solution $x(t) = x(t, 2, 4)$ of (5.31) is continuously dependent on the initial value in the sense of Definition 5.6.1.

The last example demonstrates that a careful analysis of 'closeness' of solutions must be done, if one considers equations with variable moments of impulses. We must be attentive exceptionally, if one considers continuous dependence of a solution on the right-hand side.

Beside the (5.1), consider another system

$$\begin{aligned} x' &= f(t,x) + \phi(t,x), \\ \Delta x|_{t=\tau_i(x)+\omega_i(x)} &= J_i(x) + V_i(x), \end{aligned} \tag{5.32}$$

where functions ϕ, ω, and V are all continuous.

Definition 5.6.3. The solution $x(t)$ of (5.1) continuously depends on the right-hand side of the equation if for an arbitrary $\epsilon > 0$ there exists $\delta > 0$ such that every solution $y(t) = y(t, t_0, x_0)$, of (5.32) is in the ϵ-neighborhood of $x(t)$, whenever $\|\phi(t,x)\| < \delta$, $\|V_i(x)\| < \delta$, $|\omega_i(x)| < \delta$ for all t, i, x.

Theorem 5.6.3. *Suppose (N1)–(N3),(N7)–(N9), and inequalities (5.11), are fulfilled, $k(l_f, l_J, l_\tau) < 1$, and functions g, W, ω satisfy $||g(t,x)-g(t,y)||+||W_i(x)-W_i(y)|| + |\omega_i(x) - \omega_i(y)| \le l_2||x-y||$, where l_2 is a positive constant, for all $x, y \in G$ and $(t,i) \in I \times \mathcal{A}, \omega_i(x) \ge 0$ if $i \ge j$, and $\omega_i(x) \le 0$ if $i < j$, for all $x \in G$. Then the solution $x(t)$ of (5.1) continuously depends on the right-hand side of the equation.*

Theorem 5.6.4. *Suppose (N1)–(N3),(N7)–(N9), and inequalities (5.11), are fulfilled, $k(l_f, l_J, l_\tau) < 1$, and functions g, W, ω are such that $\omega_i(x) \equiv 0, ||g(t,x) - g(t,y)|| + ||W_i(x) - W_i(y)|| \le l_2||x-y||$, where l_2 is a positive constant, for all $x, y \in G, (t,i) \in I \times \mathcal{A}$. Then each solution $x(t) = x(t,t_0,x_0), (t_0,x_0) \in I \times G$, of (5.1) continuously depends on the right-hand side of the equation on each closed finite interval, where it is defined.*

Exercise 5.6.2. Prove Theorems 5.6.3 and 5.6.4.

Remark 5.6.2. There exists another way of investigation of the continuity in parameters and the right-hand side. Indeed, if we suppose that conditions (N1)–(N3),(N7)–(N9), and inequalities (5.11) are fulfilled, and maps Π_i are homeomorphisms on G for all i, then one can prove that the continuous dependence on initial conditions of the solution $x(t)$ presents.

5.7 Generalities of Stability

Let us consider system (5.1) again, and assume that conditions (N1)–(N4) are valid, and $I = [0,\infty), \mathcal{A} = \{1,2,\ldots\}, 0 < \tau_1(x) < \tau_2(x) < \ldots$. Let $x(t)$ be a solution of (5.1) such that $x(t) : [0,\infty) \to G$. Denote $G_0 = \{(t,x) : 0 \le t \le \tau_1(x), x \in G\}$, and $G_i, i \ge 1$, has the sense assigned above, that is, $G_i = \{(t,x) : \tau_i(x) < t \le \tau_{i+1}(x), x \in G\}, i \ge 1$. Moreover, let $D(t_0,\delta) = \{(t_0,x) : \|x - x(t_0)\| < \delta\}$, where $t_0 \in I$ is fixed, be the disc with the center at $(t_0, x(t_0))$ and the radius $\delta > 0$. If $(t_0, x(t_0)) \in I \times G$, then it belongs to some $G_j, j \ge 0$. Fix this j.

Definition 5.7.1. The solution $x(t)$ is said to be B-stable if to any $\epsilon > 0$ there corresponds $\delta(t_0, \epsilon) > 0$ such that any other solution $y(t)$ is in the ϵ-neighborhood of $x(t)$ on $[t_0, \infty)$, as soon as $(t_0, y(t_0)) \in D(t_0, \delta) \cap G_j$.

Definition 5.7.2. The solution $x(t)$ is said to be uniformly B-stable if δ from Definition 5.7.1 can be chosen independently of $t_0 \in I$.

Definition 5.7.3. The solution $x(t)$ is called B-asymptotically stable, if it is B-stable, and there is $\nu > 0$ such that to any $\epsilon > 0$ there corresponds $t_1(\epsilon) > 0$ such that any other solution $y(t)$ is in the ϵ-neighborhood of $x(t)$ on $[t_0 + t_1, \infty)$, as soon as $(t_0, y(t_0)) \in D(t_0, \nu) \cap G_j$.

Definition 5.7.4. The solution $x(t)$ is called uniformly B-asymptotically stable, if it is uniformly B-stable, and the numbers ν and t_1 from the last definition can be chosen independently of $t_0 \in I$.

The reader can see that the last definitions can be easily transformed to definitions of stability for equations with fixed moments of impulses in Sect. 3.1, if $\tau_i(x) \equiv \theta_i = const, i \geq 1$. Moreover, they are essentially different from those of ordinary differential equations.

Example 5.7.1. Consider the following scalar system (scalar, since the variable x is one-dimensional, and a system, since it consists of two equations)

$$\begin{aligned} x' &= 0, \\ \Delta x|_{t=x} &= -1, \end{aligned} \tag{5.33}$$

where $t, x \in \mathbb{R}$. It has a unique surface of discontinuity $\Gamma = \{(t, x) : t = x\}$. Consider solution $x(t) = x(t, 1, 1)$. Since $(1, 1) \in \Gamma$, it starts with a jump so that $x(1+) = 0$, and $x(t) \equiv 0$ on $(1, \infty)$. Let $x_1(t) = x(t, 1, z)$, be another solution, with $z < 1$. The initial value, z, can be chosen arbitrarily close to 1, but the solution does not have a jump for $t \geq 1$, and $x_1(t) \equiv z$ on $[1, \infty)$. So, we have paradoxical situation: the closer z to 1, the larger the distance between $x(t)$ and $x_1(t)$ on $(1, \infty)$. Despite the system has "good" properties. The reason of this awkward state of deal is that $x(t)$ starts on the surface of discontinuity. Hence the initial conditions for the stability discussion should be chosen more carefully. In this example, the neighbor (t_0, y_0) must be taken such that $y_0 \geq t_0$.

Consider now, beside (5.1), the perturbed system (5.32) again.

Definition 5.7.5. The solution $x(t)$ of (5.1) is totally stable (or stable under persistent disturbances), if given $\epsilon > 0$ one can find $\delta > 0$ such that if $(t_0, x(t_0)) \notin \Gamma_i$, and $\|\phi(t, x)\| < \delta$, $\|V_i(x)\| < \delta$, $|\omega_i(x)| < \delta$ for all t, i, x, then $y(t), y(t_0) = x(t_0)$, a solution of (5.32), is in the ϵ-neighborhood of $x(t)$ on $[t_0, \infty)$.

We have special circumstances, when discussing stability of the zero solution. Assume that $f(t, 0) = 0$, $J_i(0) = 0$, for all $i \geq 1$ in (5.1), and consider the following definitions.

Definition 5.7.6. The zero solution of (5.1) is stable if to any $\epsilon > 0$ and $t_0 \in I$, there corresponds $\delta(t_0, \epsilon) > 0$ such that if $\psi(t)$ is a solution of (5.1) with $||\psi(t_0)|| < \delta(t_0, \epsilon)$, then $||\psi(t)|| < \epsilon$ for $t \geq t_0$.

Definition 5.7.7. The zero solution is uniformly stable, if the δ in the last definition is independent of t_0.

Definition 5.7.8. The zero solution is asymptotically stable if it is stable and if there exists $\nu(t_0) > 0$ such that if $\psi(t)$ is a solution of (5.1) with $||\psi(t_0)|| < \nu(t_0)$, then $\|\psi(t)\| \to 0$ as $t \to \infty$.

Definition 5.7.9. The zero solution is uniformly asymptotically stable if it is uniformly stable, and to any positive ϵ there corresponds $T(\epsilon) > 0$ such that if $\psi(t)$ is a solution of (5.1) with $\|\psi(t_0)\| < \gamma_0, \gamma_0$ is independent of t_0, then $||\psi(t)|| < \epsilon$, for all $t \geq t_0 + T(\epsilon)$.

Definition 5.7.10. The zero solution is asymptotically stable in large if it is asymptotically stable and each solution $\psi(t)$ satisfies $\psi(t) \to 0$ as $t \to \infty$.

Definition 5.7.11. The zero solution is unstable at $t_0 \in I$ if there exists a number $\epsilon_0 > 0$ such that to any $\delta > 0$ there corresponds a solution $y_\delta(t), ||y_\delta(t_0)|| < \delta$, of (5.1) such that either the solution is not continuable to ∞ or there exists a moment $t_1, t_1 > t_0$ with $||y_\delta(t_1)|| \geq \epsilon_0$.

Example 5.7.2. Prove that the zero solution of the system

$$\begin{aligned} x' &= kx, \\ \Delta x|_{t \neq i+x} &= -x^2, \end{aligned} \tag{5.34}$$

where $t, x \in \mathbb{R}, i \in \mathbb{Z}$, is asymptotically stable, if the constant k is a negative real number.

Solution. First, we construct a system, which is B-equivalent to (5.34) on $\mathbb{R}$. Fix $i \in \mathbb{Z}$. We have that $x_0(t) = x(t, i, x) = xe^{k(t-i)}$, the solution of

$$x' = kx. \tag{5.35}$$

Consider the equation $t = i + xe^{k(t-i)}$ to find the intersection moment of $x_0(t)$ with the surface $t = i + x$. It is equal to $\zeta_i = i + \phi(x)$, where $u = \phi(x)$ is an implicit function defined by the equation

$$u = xe^{ku}. \tag{5.36}$$

Thus, $x_0(\zeta_i) = xe^{k\phi(x)}$. Similarly, $x_1(t) = x(t, \zeta_i, x_0(\zeta_i) - x_0^2(\zeta_i)) = (x_0(\zeta_i) - x_0^2(\zeta_i))e^{k(t-\zeta_i)}$. Now, $W_i(x) = x_1(i) - x = -x^2e^{k\phi(x)}$, and the equivalent system has the form

$$\begin{aligned} y' &= -2y, \\ \Delta y|_{t \neq i} &= -y^2e^{k\phi(y)}. \end{aligned} \tag{5.37}$$

One can easily see that ϕ is a continuously differentiable function and, consequently, W_i are smooth functions. It is easily seen that $W_i(0) = 0$. Hence, for $h > 0$ there exists a Lipschitz constant $l(h) > 0, l(h) \to 0$, as $h \to 0$. For an arbitrary solution $y(t) = y(t, t_0, y_0)$ of (5.37) one has that

$$y(t) = e^{k(t-t_0)} y_0 + \Sigma_{t_0 \le i < t} W_i(y(i)).$$

By using the Gronwall–Bellman Lemma for discontinuous functions, we have that

$$|y(t)| \le |y_0| e^{k(t-t_0)} \Pi_{t_0 \le i < t}(1 + l(h)).$$

The last inequality implies immediately that the zero solution is asymptotically stable if $k + \ln(1 + l(h)) < 0$. The truth of the last inequality is obvious if h is sufficiently small.

5.8 B-Equivalence: Quasilinear Systems

In this section, we want to see the use of the reduction method for investigation of the following system with impulse actions:

$$\begin{aligned} \frac{dx}{dt} &= A(t)x + f(t, x), \\ \Delta x|_{t=\theta_i+\tau_i(x)} &= B_i x + I_i(x), \end{aligned} \tag{5.38}$$

where $t \in \mathbb{R}, x \in \mathbb{R}^n, \theta = \{\theta_i\}, i \in \mathbb{Z}$, is a B-sequence, the entries of the $n \times n$ matrix $A(t)$ are continuous real valued functions, B_i are real valued square matrices of order n, and $\tau_i(x)$, are positive real valued continuous functions defined on $\mathbb{R}^n, i \in \mathbb{Z}$.

We may assume additionally that:

(Q1) there exists a positive constant κ such that $\theta_{i+1} - \theta_i \ge \kappa, i \in \mathbb{Z}$;
(Q2) there exists a positive constant $l < \kappa/4$ such that for all $t \in \mathbb{R}, x, y \in \mathbb{R}^n$, $i \in \mathbb{Z}$, the following inequality is valid

$$\|f(t,x) - f(t,y)\| + \|I_i(x) - I_i(y)\| + |\tau_i(x) - \tau_i(y)| \le l\|x - y\|, \tag{5.39}$$

and $|\tau_i(x)| < l$;
(Q3) $\det(\mathcal{I} + B_i) \ne 0, i \in \mathbb{Z}$;
(Q4) for a fixed positive number H,

$$\sup_{-\infty<t<+\infty, \|x\|<H} \|f(t,x)\| + \sup_{-\infty<i<+\infty, \|x\|<H} \|I_i(x)\| = M < +\infty$$

and

$$\sup_{-\infty<t<+\infty} \|A(t)\| + \sup_{-\infty<i<+\infty} \|B_i\| = N < +\infty;$$

(Q5) $\tau_i(x) \geq \tau_i(x + I_i(x)), i \in \mathbb{Z}$, if $\|x\| \leq H$, and $l[NH + M] < 1$.

By virtue of Lemma 5.3.1 and assumption $(Q5)$ every solution $x(t), \|x(t)\| < H$, of (5.38) intersects each surface $\Gamma_i : t = \theta_i + \tau_i(x), i \in \mathbb{Z}$, at most once. Moreover, the solution intersects a surface if the range of the surface is a subset of the domain of the solution. Following the proof of Theorem 5.2.4, continuation of solutions of ordinary differential equation

$$\frac{dx}{dt} = A(t)x + f(t,x), \tag{5.40}$$

and the condition $|\theta_i| \to \infty$ as $|i| \to \infty$, one can find that every solution $x(t) = x(t, t_0, x_0), (t_0, x_0) \in \mathbb{R} \times \mathbb{R}^n$, of (5.38) is continuable on to $\mathbb{R}$, and intersect every surface of discontinuity exactly once.

We will start with a basic theorem, which is a specific form of Theorem 5.5.1 for the quasilinear case.

Fix $i \in \mathbb{Z}$. Let $x_0(t)$ be the solution of (5.40) with $x_0(\theta_i) = x$, and ξ_i the solution of the equation $t = \theta_i + \tau_i(x_0(t))$, that is, the intersection moment with Γ_i, and let $x_1(t)$ be thc solution of system (5.40) with the initial condition $x_1(\xi_i) = B_i x_0(\xi_i) + I_i(x_0(\xi_i))$. Obviously, both of these two solutions exist. Similar to W_i in (5.13) one can construct the following B-map,

$$\begin{aligned} J_i(x) = {} & (\mathcal{I} + B_i) \int_{\theta_i}^{\xi_i} (A(\tau)x(\tau) + f(\tau, x(\tau)))d\tau + \\ & I_i(x(\xi_i)) + \int_{\xi_i}^{\theta_i} (A(\tau)x_1(\tau) + f(\tau, x_1(\tau)))d\tau, \end{aligned} \tag{5.41}$$

and the system

$$\begin{aligned} \frac{dy}{dt} &= A(t)y + f(t,y), \\ \Delta y|_{t=\theta_i} &= B_i y + J_i(y). \end{aligned} \tag{5.42}$$

Exercise 5.8.1. Prove that $J_i(x) = I_i(x)$ if $\tau_i(x) = 0$, that is (θ_i, x) is a point belonging to the discontinuity surface.

Let $b(l) = l[NH + M]$, and assume that $b(l) < \min(1, H)$. Moreover, set $h = H - b(l), c(l) = b(l)(1 + N + \|\mathcal{I}\|) + M, k(l) = e^{l(N+l)}\{(1 + N)(N + l) + [(NH + M)(N + 2) + (N + l)(1 + l + b(l))e^{l(N+l)} + 1](1 - b(l))^{-1}\}$.

Theorem 5.8.1. *Assume that (5.38) satisfies conditions* $(Q1)$*–*$(Q5)$*.*
Then:

(1) functions $J_i, i \in \mathbb{Z}$*, are defined on* $\mathbb{R}^n$*;*
(2) systems (5.38) and (5.42) are B*-equivalent in* $G_H = \{x \in \mathbb{R}^n : \|x\| \leq H\}$*, such that if a solution of (5.38) or (5.42) get values in* G_h*, then the corresponding solution of (5.42)or (5.38) get values in* $G_H = \{x \in \mathbb{R}^n : \|x\| \leq H\}$*;*
(3) the following inequalities are valid

$$\|J_i(x) - J_i(y)\| \leq lk(l)\|x(\theta_i) - y(\theta_i)\|, \tag{5.43}$$

uniformly with respect to all $i \in \mathbb{Z}$*, for all* x, y *such that* $\|x(\xi_i, \theta_i, x)\| < h$ *and* $\|y(\phi_i, \theta_i, y)\| < h$*, where* $x(t, \theta_i, x)$ *and* $y(t, \theta_i, y)$ *are solutions of (5.40), and* ξ_i, ϕ_i*, are their meeting moments with the surface* Γ_i*.*

Proof. The first part of the theorem is verified above, since solutions $x_0(t)$ and $x_1(t)$ of (5.40) in (5.41) always exist.

To verify Part (2) we should remark that for a given $x \in \mathbb{R}^n$, the value $J_i(x)$ is given by $J_i(x) = x_1(\theta_i) - x = x(\theta_i, \xi_i, x_0(\xi_i)) - x$, where all considered solutions are of (5.40). Let $x(t)$ be a solution of (5.38), $\|x(t)\| < h$, and ξ_i the moment of discontinuity of this solution. Assume that $x(\theta_i) = y(\theta_i) = x_0(\theta_i)$, where $y(t)$ is a solution of (5.42). We should prove that $x(\xi_i+) = y(\xi_i) = x_1(\xi_i)$. Indeed, equality $y(\xi_i) = x_1(\xi_i)$ follows from the definition of J_i, and $x(\xi_i+) = x_0(\xi_i) + I_i(x_0(\xi_i)) = x_1(\xi_i)$ since $x(t)$ is the solution of (5.38). By employing integral equations corresponding to (5.40), we find that $\|y(t)\| < H$ if t is between θ_i and ξ_i. Since inside of intervals of continuity the solutions satisfy the same equation (5.40), one can see that the equivalence is proved in one direction. Similarly we can discuss, if begin with $y(t)$ as a solution of (5.42), $\|y(t)\| < h$. Thus, the equivalence is proved.

We next prove inequality (5.43). Let $\|x\| \leq h$. By employing integral equations corresponding to (5.40), again, we find that the solutions $x_0(t)$ and $x_1(t)$ determined above satisfy the inequalities $\|x_0(t)\| < H$ and $\|x_1(t)\| < H$ on $[\theta_i, \xi_i]$. Let $y_0(t)$ be a solution of (5.40) for which $y_0(\theta_i) = y$ and $\|y\| \leq h$. Let $\phi_i \geq \xi_i$ be a solution of the equation $t = \theta_i + \tau_i(y_0(t))$, and let $y_1(t)$ be the solution of (5.40) with the initial condition $y_1(\phi_i) = B_i y_0(\phi_i) + I_i(y_0(\phi_i))$. We have that

$$x_0(t) = x + \int_{\theta_i}^{t} [A(s)x_0(s) + f(s, x_0(s))]ds,$$

$$y_0(t) = y + \int_{\theta_i}^{t} [A(s)y_0(s) + f(s, y_0(s))]ds.$$

Subtract the second equation from the first one, and apply the Lipschitz condition and the Gronwall–Bellmann Lemma to obtain

$$\|x_0(t) - y_0(t)\| \leq e^{(N+l)l}\|x - y\|, t \in [\xi_i, \phi_i]. \tag{5.44}$$

Next, one can write that

$$\phi_i - \xi_i = \tau_i(y_0(\phi_i)) - \tau_i(x_0(\xi_i)) \le l\|y_0(\phi_i) - x_0(\xi_i)\| \le$$

$$l[\|y_0(\phi_i) - x_0(\phi_i)\| + \|x_0(\phi_i) - x_0(\xi_i)\|] \le le^{(N+l)l}\|x - y\| +$$

$$l\|\int_{\xi_i}^{\phi_i}[A(s)y_0(s) + f(s, y_0(s))]ds\| \le le^{(N+l)l}\|x - y\| + b(l)(\phi_i - \xi_i).$$

The last inequality implies

$$\phi_i - \xi_i \le \frac{le^{(N+l)l}}{1 - b(l)}\|x - y\|. \tag{5.45}$$

Now, we have

$$x_1(\xi_i+) - y_1(\xi_i) = x_0(\xi_i) + I_i(x_0(\xi_i)) - y_0(\xi_i) - \int_{\xi_i}^{\phi_i}[A(s)y_0(s)$$
$$+ f(s, y_0(s))]ds - I_i(y_0(\xi_i) +$$
$$\int_{\xi_i}^{\phi_i}[A(s)y_0(s) + f(s, y_0(s))]ds) + \int_{\xi_i}^{\phi_i}[A(s)y_1(s) + f(s, y_1(s))]ds,$$

and

$$\|x_1(\xi_i+) - y_1(\xi_i)\| \le \frac{(1 + l + b(l))e^{(N+l)l}}{1 - b(l)}\|x - y\|.$$

Consequently,

$$\|x_1(t) - y_1(t)\| \le \frac{(1 + l + b(l))e^{2(N+l)}}{1 - b(l)}\|x - y\|. \tag{5.46}$$

Finally, subtracting the expression

$$J_i(y) = (\mathcal{I} + B_i)\int_{\theta_i}^{\phi_i}(A(\tau)y_0(\tau) + f(\tau, y_0(\tau)))d\tau + I_i(y_0(\phi_i)) +$$

$$\int_{\phi_i}^{\theta_i}(A(\tau)y_1(\tau) + f(\tau, y_1(\tau)))d\tau$$

from (5.41) and using (5.44), (5.45), and (5.46), we conclude that (5.43) holds. The theorem is proved. □

Bounded Solutions. Using the transformation $y = U(t)z$, which has been defined in Sect. 4.1, such that matrices $U(t), U^{-1}(t)$, are bounded on $\mathbb{R}$, one can reduce (5.42) to the system

$$\frac{d\xi}{dt} = P_1(t)\xi + f_1(t,z),$$
$$\Delta\xi|_{t=\theta_i} = Q_i^1\xi + I_i^1(z),$$
$$\frac{d\eta}{dt} = P_2(t)\eta + f_2(t,z),$$
$$\Delta\eta|_{t=\theta_i} = Q_i^2\eta + I_i^2(z), \tag{5.47}$$

where $z = (\xi,\eta)$, $\xi \in \mathbb{R}^m$, $\eta \in \mathbb{R}^{n-m}$.

In what follows, without loss of generality, we assume that the system (5.42) has the form (5.47).

Theorem 5.8.2. *Assume that conditions $(Q1)$–$(Q5)$ are satisfied, associated with (5.38) system (4.1) is exponentially dichotomous and*

(Q6) $2KM[\frac{1}{\gamma} + \frac{c(l)e^{2\gamma\kappa}}{1-e^{-\frac{\gamma\kappa}{2}}}] + b(l) < H;$

(Q7) $2Kl[\frac{1}{\gamma} + \frac{k(l)e^{2\gamma\kappa}}{1-e^{-\frac{\gamma\kappa}{2}}}] < 1.$

Then (5.38) has a unique solution bounded on $\mathbb{R}$.

Proof. Consider the following system of integral equations:

$$\xi(t) = \int_{-\infty}^{t} X_1(t,\tau)f_1(\tau,z)d\tau + \sum_{\theta_i<t} X_1(t,\theta_i+)J_i^1(z),$$
$$\eta(t) = -\int_{t}^{\infty} X_2(t,\tau)f_2(\tau,z)d\tau - \sum_{\theta_i\ge t} X_2(t,\theta_i+)J_i^2(z), \tag{5.48}$$

and the sequence of approximations $z_k = (\xi_k,\eta_k), k \ge 0, \xi_0 \equiv 0, \eta_0 \equiv 0,$

$$\xi_{k+1}(t) = \int_{-\infty}^{t} X_1(t,\tau)f_1(\tau,z_k)d\tau + \sum_{\theta_i<t} X_1(t,\theta_i+)J_i^1(z_k),$$

$$\eta_{k+1}(t) = -\int_{t}^{\infty} X_2(t,\tau)f_2(\tau,z_k)d\tau - \sum_{\theta_i\ge t} X_2(t,\theta_i+)J_i^2(z_k).$$

Using $(Q6)$ one can check that all approximations satisfy $\|z_k(t)\| < h$, if $t \in \mathbb{R}$. Indeed, we have that $\|z_0\| = 0, t \in \mathbb{R}$. Assume that $\|z_k\| < h$. Then applying (4.31), we have

$$\|\xi_{k+1}(t)\| \le \int_{-\infty}^{t} Ke^{-\gamma(t-\tau)}Md\tau + \sum_{t_i<t} Ke^{-\gamma(t-\theta_i)}M < KM[\frac{1}{\gamma} + \frac{c(l)e^{2\gamma\kappa}}{1-e^{-\frac{\gamma\kappa}{2}}}].$$

Similar evaluation can be made for $\eta_{k+1}(t)$. So, finally, we have that $\|z_{k+1}(t)\| \le 2K[\frac{M}{\gamma} + \frac{c(l)}{1-e^{-\frac{\gamma\kappa}{2}}}] < H - b(l) = h$. That is, $\|z_k(t)\| < h, k \ge 0$, if $t \in \mathbb{R}$, by the induction.

Analogously, condition $(Q7)$ implies that the sequence z_k is convergent over $\mathbb{R}$. Thus, the integral equation has a solution $z_0(t)$, with $\|z_0(t)\| \leq h$ for $t \in \mathbb{R}$. The uniqueness can be verified easily. The theorem is proved. □

Exercise 5.8.2. Explain, why we do not use an analogue of condition (N6) in the last theorem. Discuss uniqueness of solutions of system (5.38), and the uniqueness of the solution bounded on $\mathbb{R}$.

5.9 Poincaré Criterion and Periodic Solutions of Quasilinear Systems

Theorem 5.9.1. *Assume that conditions (M0),(N1)–(N4) are fulfilled, (5.1) is an* (ω, p)*-periodic system, there exist a moment* $t_0 \in \mathbb{R}$ *and a solution* $x(t)$ *of (5.1) with* $x(t_0+\omega) = x(t_0)$. *Then there exists an* ω*-periodic solution of the system.*

Proof. The necessity can be verified easily: if $x(t)$ is an ω-periodic solution, then $x(t_0+\omega) = x(t_0)$ for all $t_0 \in \mathbb{R}$. Let us prove sufficiency. Assume that $x(t_0+\omega) = x(t_0)$ is true for a fixed $t_0 \in \mathbb{R}$ and a solution $x(t)$. Without loss of generality, assume that t_0 is not a point of discontinuity of $x(t)$, and (t_0, x_0) lies between two consecutive surfaces Γ_i and Γ_{i+1}. Then $(t_0+\omega, x_0)$ is between Γ_{i+p} and Γ_{i+1+p}. Consider $t \geq t_0$. The theorem on periodic systems of ordinary differential equations [59] implies that $x(t+\omega)$ is a solution of (5.2), and $x(t) = x(t+\omega)$ near t_0 while both solutions are continuous. More precisely, solutions of (5.2), which represent $x(t)$ and $x(t+\omega)$ on the intervals of continuity are equal to each other. The nearest moments of discontinuity from the right are defined by equations $t = \tau_{i+1}(x(t))$ and $t = \tau_{i+1+p}(x(t+\omega))$. The first of these equations has the solution $t = \theta_i$. Denote the solution of the second one as $t = \theta'_{i+1+p}$. We have that $\theta'_{i+1+p} = \tau_{i+1+p}(x(\theta'_{i+1+p}+\omega)) = \tau_{i+1+p}(x(\theta'_{i+1+p})) = \tau_{i+1}(x(\theta'_{i+1+p})) + \omega$. The absence of beating and $x(t) = x(t+\omega)$ imply that $\theta'_{i+1+p} = \theta_{i+1+p}$ and $\theta_{i+1+p} = \theta_{i+1}+\omega$. Next, we have that $x(\theta_{i+1}+) = J_{i+1}(x(\theta_{i+1+p})) = J_{i+1+p}(x(\theta_{i+1+p}+\omega)) = x(\theta_{i+1+p}+\omega+)$. Consequently, the two solutions are equal to each other while both solutions are continuous, near $t = \theta_{i+1+p}$, for increasing t, and $x(t+\omega)$ is a solution of (5.1) for these values of the argument. One can proceed in this way such that, finally, we obtain that $x(t+\omega)$ is a solution of (5.1), and, moreover, $x(t+\omega) = x(t)$, if $t \in [t_0, t_0+\omega]$. The assertion is proved. □

Suppose that (5.38) is an (ω, p)-periodic system, i.e., $A(t)$ and $f(t,x)$ are ω-periodic functions of t, there exists an integer p related to ω by the condition $\theta_{i+p} = \theta_i + \omega$, $i \in \mathbb{Z}$, and $B_{i+p} = B_i$, $\tau_{i+p}(x) = \tau_i(x)$ for all $i \in \mathbb{Z}, x \in \mathbb{R}^n$. Since (5.38) has the uniqueness property, it can be shown that the following result holds.

Lemma 5.9.1. *Assume that (5.38) is an* (ω, p)*-periodic system. Then the sequence of maps* $J_i(x)$ *is* p*-periodic in* i, *and (5.42) is an* (ω, p)*-periodic system.*

Exercise 5.9.1. Prove Lemma 5.9.1.

Hint: Use the technique of the last proof.

Theorem 5.9.2. *If conditions* $(Q1)$*–*$(Q7)$ *are satisfied, and associated with (5.38) system (4.3) is exponentially dichotomous, then the* (ω, p)*-periodic system (5.38) has a unique* ω*-periodic solution.*

Proof. By Theorem 5.8.2 system (5.47) has a unique bounded solution $z^0(t)$, which satisfies (5.48). Let us prove that it is ω-periodic. Following the proof of Theorem 5.8.2, we only need to show that the approximations are ω-periodic functions. Let us apply the induction method. We have that $z_0(t) \equiv 0$. Assume that the approximation $z_k(t)$ is ω-periodic function. Consider the approximation z_{k+1}. We will check the periodicity of ξ_{k+1}, since that for the second component η_{k+1} is very similar.

$$\xi_{k+1}(t+\omega) = \int_{-\infty}^{t+\omega} X_1(t+\omega,\tau) f_1(\tau, z_k(\tau))d\tau$$

$$+ \sum_{t_i < t+\omega} X_1(t+\omega, \theta_i) J_i^1(z_k(\theta_i)) =$$

$$\int_{-\infty}^{t} X_1(t+\omega, \tau+\omega) f_1(\tau+\omega, z_k(\tau+\omega))d\tau$$

$$+ \sum_{t_i < t} X_1(t+\omega, \theta_i+\omega) J_{i+p}^1(z_k(\theta_i+\omega)) =$$

$$\int_{-\infty}^{t} X_1(t,\tau) f_1(\tau, z_k(\tau))d\tau + \sum_{t_i < t} X_1(t,\theta_i) J_i^1(z_k(\theta_i)) = \xi_k(t)$$

The theorem is proved. □

Let us denote $\bar{M} = \max_{t,s\in[0,\omega]} ||G(t,s)||$, where $G(t,s)$ is the Green's function (4.66). The following theorem is valid.

Theorem 5.9.3. *Suppose conditions (Q1)–(Q5) are satisfied, associated system (4.1) is exponentially dichotomous and*

(Q8) $\bar{M}M(T+p) < h$;
(Q9) $\bar{M}l(1+lk(l)) < 1$.

Then (5.38) has a unique ω*-periodic solution.*

Exercise 5.9.2. Prove Theorem 5.9.3.

Notes

The problem of investigation of differential equations with solutions, which have discontinuities on surfaces placed in (t,x)-space is one of the most difficult and interesting subjects of the theory [2,4,14,20,32–36,69,82,85,95–97,102,138,141,

142, 144, 145, 153, 161]. It was emphasized in early stage of theory's development in [111]. Conditions of "absence of beating" of solutions against surfaces of discontinuity were firstly defined in [138]. The theoretical importance caused further research of the beating phenomena [63, 82]. Another complexity in the analysis of the systems, except the beating of solutions, is description of the closeness of solutions with different moments of discontinuity. The problem was considered earlier in theory of functions [90, 147]. In implicit form, the description was done for impulsive differential equations in three cases: for stability of solutions of equations with variable moments of impulses; continuous dependence of solutions on parameters, when moments of impulses are not fixed; analysis of almost periodicity of discontinuous solutions in both cases of systems with fixed and nonfixed moments of impulses [30, 75, 95, 97, 138, 141, 142] . By using basic ideas of our predecessors [75, 141, 147], we introduce a special topology [2–4, 4–7, 7, 8, 32, 33, 35, 36] in a set of piecewise continuous functions having, in general, points of discontinuity, which do not coincide. Thus, we operate with the concepts of B-topology, B-equivalence, and ϵ-neighborhoods, when we investigate systems with variable moments of impulses [2–8, 32, 33, 35, 36] or consider almost periodicity of solutions of impulsive systems [7, 31]. In [90], it was explored that a topology in spaces of piecewise continuous functions can be metricized. We specify this result in [31] for investigation of almost periodic discontinuous nonautonomous systems. It is not surprising that the topology begins to be useful for other differential equations with discontinuities of different types [7, 17, 20, 27]: Filippov's type differential equations; differential equations on variable time scales. The most important concept used in the chapter is the method of reduction to systems with fixed moments of impulses, i.e., the B-equivalence method. It was introduced and developed in [1–4, 13, 15, 20, 21, 25–32]. The material of the chapter lies fully on results obtained by B-equivalence method. One must say that some of these results are published for the first time in the present book. Let us list these results: condition (N3) as a general source of the absence of beating; content of Sects. 5.5, 5.7, 5.9; Theorems 5.2.1–5.2.4 of existence and uniqueness; condition (N5), which provides uniqueness to the left extension, and Theorem 5.3.4, which guarantees the condition. Condition (N6) cancels many difficulties of the left continuation. Definitions 5.7.1–5.7.4 of stability for differential equations, where the importance of discs $D(t_0, \delta)$ is emphasized, are newly given. Lemmas 5.3.1, 5.3.2 and Theorems 5.3.1, 5.3.2 are due to [28]. Results of Sect. 5.8 are published in [33]. Let us point out that the B-equivalence method is effective not only in bounded domains but it can also be applied successfully if impulsive equations are considered with unbounded domains [37]. Exceptionally, it is important for existence of global manifolds. Linearization in the neighborhood of the nontrivial solution, the central auxiliary result of the stability theory is solved in [2]. The problem of controllability of boundary-value problems for quasilinear impulsive system of integro–differential equations is investigated in [18]. Finally, the method also proves its effectiveness to indicate chaos and shadowing property of impulsive systems [9, 11].

Chapter 6
Differentiability Properties of Nonautonomous Systems

In this chapter, we investigate the fundamental properties of differential equations with variable moments of impulses: differential and analytic dependence of solutions on initial conditions and parameters. Differentiability of solutions is the property, which is of underestimated importance for differential equations. One needs the conditions, which provide the smoothness of solutions if a system is to be linearized around a certain solution, to prove the existence of periodic and almost periodic solutions in critical and noncritical cases by using the method of small parameter [105, 107], to investigate problems of synchronization and bifurcation theory.

We consider the following system of impulsive differential equations:

$$\begin{aligned} x' &= f(t, x), \\ \Delta x|_{t=\tau_i(x)} &= J_i(x). \end{aligned} \tag{6.1}$$

The system is defined on the set $\Omega = I \times \mathcal{A} \times G$ of points (t, i, x), where $G \subset \mathbb{R}^n$ is an open and connected set, I an open interval of $\mathbb{R}$, and $\mathcal{A}$ an interval of $\mathbb{Z}$. We assume that $f(t, x)$ is a continuous function, J_i are functions on G, and $\tau_i(x)$ are continuous functions on G, $i \in \mathcal{A}$. Condition (M0), Chap. 2, and conditions (N1)–(N6) of Chap. 5 are valid. Moreover, we assume that

(N11) the derivatives $\partial f(t, x)/\partial x_j$, $\partial J_i(x)/\partial x_j$, $\partial \tau_i(x)/\partial x_j$, are continuous on G, uniformly for all $t \in I, i \in \mathcal{A}$.

It is required that the vector functions f, J, x, and their derivatives are column-vectors, and the derivatives of the functions τ are assumed to be vector-rows. Products of vectors and matrices are the products of rectangle matrices. The following condition will be needed throughout this chapter:

(N12) the inequality $\tau_{ix}(x) f(\tau_i(x), x) \neq 1$, is fulfilled for all $(i, x) \in \mathcal{A} \times G$.

The condition means that each solution of (6.1) may meet a surface of discontinuity only transversally.

Results of the preceding chapter imply that if $(t_0, x_0) \in I \times G$, then there exists a unique solution $x(t) = x(t, t_0, x_0), x_0 = (x_0^1, \ldots, x_0^n)$, of (6.1) on some interval $[t_0, T], T > t_0$, with points of discontinuity $t = \theta_i, i \in \mathcal{A}$. We will discuss

M. Akhmet, *Principles of Discontinuous Dynamical Systems*,
DOI 10.1007/978-1-4419-6581-3_6,

differentiability properties assuming that $t \geq t_0$, since it is sufficient for application needs. For the sake of simplicity, we denote the moments $t_0 \leq \theta_m < \ldots < \theta_k < T$, and assume that $(t_0, x_0) \in G_{m-1} = \{(t, x) : \tau_{m-1}(x) < t \leq \tau_m(x), x \in G\}$.

6.1 Differentiability with Respect to Initial Conditions

Consider the disc $D(t_0, \delta) = \{(t_0, x) : \|x - x_0\| < \delta\}$ with the center at (t_0, x_0) and with the radius $\delta > 0$, and the interval $d(x_0, \delta) = \{(t, x_0) : t_0 - \delta < t < t_0 + \delta\}$ with the center at (t_0, x_0) and with the radius $\delta > 0$.

Denote by $x^j(t), j = 1, 2, \ldots, n$, the solution of (6.1) with $x^j(t_0) = (x_0^1, \ldots, x_0^j + \xi, \ldots, x_0^n)$, and by $\eta_i^j, i = 1, 2, \ldots, k$, the points of discontinuity of this solution.

Definition 6.1.1. The solution $x(t)$ is B-differentiable with respect to $x_0^j, j = 1, 2, \ldots, n$, on $[t_0, T]$ if there exists $\delta > 0$, such that if $(t_0, x^j(t_0)) \in D(t_0, \delta) \cap G_{m-1}$, then:

(1) there exist real constants $\nu_{ij}, i = 1, 2, \ldots, k$, such that

$$\theta_i - \eta_i^j = \nu_{ij}\xi + o(|\xi|); \tag{6.2}$$

(2) for all $t \notin \widehat{(\theta_i, \eta_i^j]}, i = 1, 2, \ldots, k$, it is true that

$$x^j(t) - x(t) = u_j(t)\xi + o(|\xi|), \tag{6.3}$$

where function $u_j(t) \in \mathcal{PC}([t_0, T], \theta)$.

The pair $\{u_j(t), \{\nu_{ij}\}\}$, which consists of the function u_j and the sequence $\{\nu_{ij}\}$, is called a B-derivative of the solution $x(t)$ with respect to x_0^j.

In a similar manner, we shall define B-derivatives with respect to t_0. Denote by $x^0(t)$ a solution of (5.45) such that $x^0(t_0 + \xi) = x_0$, where ξ is a fixed real number. If $|\xi|$ is sufficiently small then $x^0(t)$ exists on $[t_0 + \xi, T]$. Denote by $\eta_i^0, i = 1, 2, \ldots, k$, the points of discontinuity of this solution in this interval.

Definition 6.1.2. The solution $x(t)$ is B-differentiable with respect to t_0 if there exists $\delta > 0$, such that if $(t_0 + \xi, x_0) \in d(x_0, \delta) \cap G_m$, then:

(1) there exist real constants $\nu_{i0}, i = 1, 2, \ldots, k$, which satisfy

$$\theta_i - \eta_i^0 = \nu_{i0}\xi + o(|\xi|); \tag{6.4}$$

(2) for all $t \notin \widehat{(\theta_i, \eta_i^0]}, i = 1, 2, \ldots, k$, it is true that

$$x^0(t) - x(t) = u_0(t)\xi + o(|\xi|), \tag{6.5}$$

where $u_0(t) \in \mathcal{PC}([t_0, T], \theta)$, and $\theta = \{\theta_i\}_{i=1,2,\ldots,k}$.

The pair $\{u_0(t), \{\nu_{i0}\}\}$, which consists of the function u_0 and the sequence $\{\nu_{i0}\}$, is called a B-derivative of the solution $x(t)$ with respect to t_0.

In the sequel, we shall write values of functions and their derivatives at points $(\theta_i, x(\theta_i))$ and $(\theta_i, x(\theta_i+))$ without mentioning the values of the arguments, and distinct the second case by the subscript $+$. To denote derivatives with respect to x and t we apply subscripts. Let us start with some auxiliary maps and results. Consider the system of ordinary differential equations

$$x' = f(t, x), \tag{6.6}$$

which is the part of (6.1), and maps $J, \tau : G \to \mathbb{R}^n, \tau : G \to \mathbb{R}$, being continuously differentiable in x. Choose a point $(\kappa, x) \in I \times G$, and keep κ fixed next. Let $t = \theta, \theta = \theta(x)$, be a moment of meeting of $x(t) = x(t, \kappa, x)$, the solution of (6.6), with the surface $t = \tau(x)$ transversally. Moreover, assume that the solution $\bar{x}(t) = x(t, \theta, x(\theta) + I(x(\theta))$ of (6.6) exists on $\widehat{[\theta, \kappa]}$. Define the B-map $W : x \to \bar{x}(\kappa)$ such that

$$W(x) = \int_\kappa^\theta f(u, x(u))du + J(x + \int_\kappa^\theta f(u, x(u))du) + \int_\theta^\kappa f(u, \bar{x}(u))du. \tag{6.7}$$

Denote by $\tilde{G} \subset G$, a domain of the map. Set $A(t) = f_x(t, x(t))$, and let $U(t), U(\kappa) = \mathcal{I}$, be the fundamental matrix of the system $u' = A(t)u$.

Lemma 6.1.1. *$\theta(x) \in C^{(1)}(\tilde{G})$, and*

$$\theta'(x) = \frac{\tau_x(x(\theta))U(\theta)}{(1 - \tau_x(x(\theta))f(\theta, x(\theta))}. \tag{6.8}$$

Proof. The solution $x(t)$ meets the surface $t = \tau(x)$ transversally at a moment $t = \theta$, that is, $\tau_x(x(\theta))f(\theta, x(\theta)) \neq 1$. Obviously, the inequality is valid for all x in a neighborhood of $x(\theta)$. By conditions of the lemma, the solution $x(t)$ is differentiable in the initial value x. The moment of intersection with $t = \theta(x)$ satisfies $t = \tau(x(t))$. More precisely, $t = \tau(x(t, \kappa, x))$. Now, we use the implicit function theorem in a neighborhood of $(\theta, x(\theta))$. We have that

$$\frac{d\theta(x)}{dx} = -\frac{-\frac{d\tau(z)}{dz}\frac{\partial x(t,\kappa,x)}{\partial x}}{1 - \frac{d\tau(z)}{dz}\frac{dx(t,\kappa,x)}{dt}}\Big|_{t=\theta, z=x(\theta)} = \frac{\tau_x(x(\theta))U(\theta)}{(1 - \tau_x(x(\theta))f(\theta, x(\theta))}.$$

The differentiability of $\theta(x)$ implies that it is continuous at x. Consequently, the last formula and continuity of functions τ_x, f, U complete the proof of the lemma. □

Corollary 6.1.1. *$\theta(x)$ is a continuous function.*

Lemma 6.1.2. $W(x) \in C^{(1)}(\tilde{G})$.

Proof. By continuous differentiability of functions involved in (6.7) and the last lemma, one can find that

$$W_x(x) = (f(\theta, x(\theta)) - f(\theta, x(\theta) + J(x(\theta)))\theta_x(x) + J_x(\mathcal{I} + f(\theta, x(\theta))\theta_x), \tag{6.9}$$

and W_x is a continuous matrix. The lemma is proved. □

Corollary 6.1.2. *$W(x)$ is a continuous function.*

Assume that there is an integer $r > 1$ such that the functions f, τ, and J are r times continuously differentiable in x, and the function f is $r - 1$ times continuously differentiable in t in a h-neighborhood of the surface $t = \tau(x)$, where h is a positive number.

The next lemma can be proved easily just by several differentiation of the expression in the right-hand side of (6.8), but we prefer another way, which can enrich our technique of investigation.

Lemma 6.1.3. $\theta(x) \in C^{(r)}(\tilde{G})$.

Proof. Beside the solution $x(t) = x(t, \kappa, x)$, let us consider a solution $\bar{x}(t) = x(t, \kappa, x + \Delta x)$ of (6.6). Denote by $t = \bar{\theta} = \theta(x + \Delta x)$ the meeting moment of this solution with the surface $t = \tau(x)$. Without loss of generality, assume that $\bar{\theta} \geq \theta \geq \kappa$.

We shall prove the lemma if the equality

$$\theta(x + \Delta x) - \theta(x) = \sum_{i=0}^{n} \phi_{1i}(x)\xi_i + \ldots + \sum_{i,j,\ldots s=0}^{n} \phi_{rij\ldots s}(x)\xi_i\xi_j\ldots\xi_s + o(||\Delta x||^r), \tag{6.10}$$

where coefficients ϕ are continuous and symmetric with respect to permutation of indices, will be verified. Let us denote $Q_r(\cdot)$- an r-degree polynomial with respect to a vector $(\cdot)$. Applying the differentiability of $\tau(x)$, we can find that

$$\theta(x + \Delta x) - \theta(x) = Q_r(x(\bar{\theta}, \kappa, x + \Delta x) - x(\theta)) + o(||x(\bar{\theta}, \kappa, x + \Delta x) - x(\theta)||^r). \tag{6.11}$$

Moreover, differentiability of f, and consequently, smoothness of solutions of (5.2) imply that

$$\int_{\theta}^{\bar{\theta}} f(u, x(u, \kappa, x + \Delta x))du = Q_r(\bar{\theta} - \theta, \Delta x) + o(||\bar{\theta} - \theta, \Delta x||^r). \tag{6.12}$$

Now, applying equality

$$x(\bar{\theta}, \kappa, x + \Delta x) - x(\theta) = \int_{\theta}^{\bar{\theta}} f(u, x(u, \kappa, x + \Delta x))du + x(\theta, \kappa, x + \Delta x) - x(\theta)$$

we obtain that

$$\bar{\theta} - \theta = \sum_{i=0}^{n} \psi_{1i}(x(\theta))\{f(\theta, x(\theta))\}_i(\bar{\theta} - \theta) +$$
$$\sum_{i=0}^{n} L_i(\Delta x)(\bar{\theta} - \theta)^i + o(||\xi||^r), \tag{6.13}$$

where $\{a\}_i$ means the i-th coordinate of a vector $a = (a_1, a_2, \ldots, a_n), L_i, i = 1, 2, \ldots, n$, are polynomials and $L_1 = o(||\Delta x||)$. The result of Lemma 6.1.1 is the relation

$$\bar{\theta} - \theta = \sum_{i=0}^{n} \phi_{1i}(x)\xi_i + o(||\Delta x||).$$

It implies that the inequality

$$1 - \sum_{i=0}^{n} \psi_{1i}(x(\theta))\{f(\theta, x(\theta))\}_i - L_1(\Delta x) \neq 0$$

is true if $||\Delta x||$ is sufficiently small. Hence,

$$\bar{\theta} - \theta = \sum_{i=0}^{n} \phi_{1i}(x)\xi_i + \sum_{i,j=0}^{n} \phi_{2ij}(x)\xi_i\xi_j + o(||\Delta x||^2). \tag{6.14}$$

Substitute the last expression in (6.13) and proceed the procedure to obtain (6.10).
The lemma is proved. □

Lemma 6.1.4. $W(x) \in C^{(r)}(\tilde{G})$.

Proof. The last lemma imply that expression (6.9) is $r - 1$ times differentiable. The lemma is proved. □

To investigate differentiability of solutions, we begin with the following auxiliary impulsive system of differential equations

$$\frac{dy}{dt} = f(t, y),$$
$$\Delta y|_{t=\theta_i} = W_i(y), \tag{6.15}$$

where moments of impulses θ_i are fixed, and W_i are functions defined on G. Assume that a solution $y(t) = y(t, t_0, x_0)$ of this system exists and is defined on the interval $[t_0, T]$.

Lemma 6.1.5. *If functions f and W_i are continuous and have continuous partial derivatives in coordinates of $y \in G$, then the solution $y(t) = y(t, t_0, x_0)$ of (6.15) is B-differentiable with respect to initial conditions. The first components of the B-derivatives, $u_j, j = 0, 1, 2, \ldots, n$, are solutions of the linear impulsive system*

$$\begin{aligned} \frac{du}{dt} &= f_y(t, y(t))u, \\ \Delta u|_{t=\theta_i} &= W_{iy}(y(\theta_i))u, \end{aligned} \tag{6.16}$$

with initial values $-f(t_0, x_0)$ if $j = 0$, and $e_j = (\underbrace{0, ..., 0, 1}_{j}, 0, ..., 0), j = 1, 2, \ldots, n$. Constants v_{ij} in (6.2) and (6.4) are zeros for all i and j.

Proof. Let us prove the theorem for x_0^1. For all other $x_0^j, j = 2, \ldots, n$, and t_0 the proof is similar.

Let $u_1(t), u_1(t_0) = e_1$, be the solution of (6.16). By the theorem of differentiability [77], we have that $y(t, t_0, x_0 + \xi e_1) - y(t, t_0, x_0) = u_1(t)\xi + p_1(\xi)$, where $p_1(\xi) = o(|\xi|)$. Let m be one of the numbers $1, 2, \ldots, k$. We assume that $y(\theta_m, t_0, x_0 + \xi e_1) - y(\theta_m, t_0, x_0) = u_1(\theta_m)\xi + p_m(\xi), p_m = o(|\xi|)$, and will show that

$$y(\theta_m+, t_0, x_0 + \xi e_1) - y(\theta_m+, t_0, x_0) = u_1(\theta_m+)\xi + \bar{p}_m(\xi), \tag{6.17}$$

where $\bar{p}_m = o(|\xi|)$. Indeed, the conditions of the lemma imply that

$$\begin{aligned} &y(\theta_m+, t_0, x_0 + \xi e_1) - y(\theta_m+, t_0, x_0) = y(\theta_m, t_0, x_0 + \xi e_1) - \\ &y(\theta_m, t_0, x_0) + W_m(y(\theta_m, t_0, x_0 + \xi e_1)) - W_m(y(\theta_m, t_0, x_0)) = \\ &u_1(\theta_m)\xi + p_m(\xi) + W_{mx}(y(\theta_m, t_0, x_0))(y(\theta_m, t_0, x_0 + \xi e_1) - \\ &y(\theta_m, t_0, x_0)) + \tilde{p}(\xi) = (\mathcal{I} + W_{mx}(y(\theta_m, t_0, x_0)))y(\theta_m, t_0, x_0)\xi + \\ &p_m(\xi) + W_{mx}(y(\theta_m, t_0, x_0))p_m(\xi) + \tilde{p}(\xi). \end{aligned} \tag{6.18}$$

Then, as $\tilde{p} = o(|\xi|)$, the formula (6.17) is true. Denote by $U(t), U(\theta_m) = \mathcal{I}$, the fundamental matrix of (6.16) and use (6.18) to obtain that for all $t \in (\theta_m, \theta_{m+1}]$

$$\begin{aligned} &y(t, t_0, x_0 + \xi e_1) - y(t, t_0, x_0) = U(t)[y(\theta_m+, t_0, x_0 + \xi e_1) - y(\theta_m+, t_0, x_0)] \\ &+p(u_1(\theta_m+)\xi + p_{m+1}(\xi)) = U(t)y(\theta_m+, t_0, x_0) + U(t)\bar{p}_{m+1}(\xi)) \\ &+p(u_1(\theta_m+)\xi + p_{m+1}(\xi)) = u_1(t)\xi + p_{m+1}(\xi), \end{aligned}$$

where $p_{m+1} = o(|\xi|)$. Thus, we obtain that

$$y(t, t_0, x_0 + \xi e_1) - y(t, t_0, x_0) = u_1(t)\xi + p(\xi), p(\xi) = o(|\xi|).$$

The lemma is proved. □

Lemma 6.1.6. *If (N1)–(N6),(N11),(N12) are valid, then the solution* $x(t) = x(t, t_0, x_0)$ *of (6.1) is continuous in initial conditions in B-topology on* $[t_0, T]$.

Proof. We consider the continuity in x_0. The dependence on t_0 can be considered similarly. Denote by $\bar{x}(t) = x(t, t_0, x_0 + \Delta x)$, another solution of the equation. Our aim is to show that for an arbitrary $\epsilon > 0$ one can find $\delta > 0$ such that $||\Delta x|| < \delta$ implies $\bar{x}(t)$ is in the ϵ-neighborhood of $x(t)$ in B-topology on $[t_0, T]$. Fix $\epsilon > 0$. For a positive number $\alpha \in R$ we shall construct a set G^α in the following way. Let $F_\alpha = \{(t, x)|t \in [t_0, T], \|x - x(t)\| < \alpha\}$, and $G_i(\alpha), i = 0, 1, 2, \ldots, k+1$, be α-neighborhoods of points (t_0, x_0), $(\theta_i, x(\theta_i))$, $i = 1, 2, \ldots, k$, $(T, x(T))$ in $\mathbb{R} \times \mathbb{R}^n$ respectively, and $\bar{G}_i(\alpha)$, $i = 1, 2, \ldots, k$, be α-neighborhoods of points $\left(\theta_i, x^0(\theta_i+)\right), i = 1, 2, \ldots, k$, respectively. Let

$$G^\alpha = F_\alpha \cup \left(\cup_{i=0}^{k+1} G_i(\alpha)\right) \cup \left(\cup_{i=1}^{k} \bar{G}_i(\alpha)\right).$$

Take $\alpha = h$ sufficiently small that sets $G^h \subset I \times G$ and $\bar{G}_i(h)$, $i = 1, 2, \ldots, k$, do not intersect any surface of discontinuity, except Γ_i. Fix $\epsilon, 0 < \epsilon < h$.

1. In view of the theorem on continuity of solutions [77], there exists $\bar{\delta}_k \in \mathbb{R}$, $0 < \bar{\delta}_k < \epsilon$, such that every solution $x_k(t)$ of (6.6), which starts in $\bar{G}_k(\bar{\delta}_k)$, is continuable to $t = T$, does not intersect any surface of discontinuity, except Γ_k, and
$$\|x_k(t) - x(t)\| < \epsilon.$$
2. The continuity of J and condition (N3) imply that there exists $\delta_k \in \mathbb{R}$, $0 < \delta_k < \epsilon$, such that $(\kappa, x) \in G_k(\delta_k)$ implies $(\kappa, x + J(x)) \in \bar{G}_k(\bar{\delta}_k)$.
3. Using continuity of solutions and condition (N3), one can find $\bar{\delta}_{k-1}$, $0 < \bar{\delta}_{k-1} < \epsilon$, such that a solution $x_{k-1}(t)$ of (6.6), which starts in $\bar{G}_{k-1}(\bar{\delta}_{k-1})$, intersects Γ_k in $G_k(\delta_k)$ (we continue the solution $x_{k-1}(t)$ only to the moment of the intersection) and $\|x_{k-1}(t) - x(t)\| < \epsilon$ for all t from the common domain of $x_{k-1}(t)$ and $x(t)$.

Continue the process for $k-2, k-3, \ldots, 1, 0$, to obtain a sequence of families of solutions of (6.6) $x_i(t), i = 1, 2, \ldots, m$, and a number $\delta \in \mathbb{R}, 0 < \delta < \epsilon$, such that a solution $\bar{x}(t) = x(t, t_0, x_0 + \Delta x)$, which starts in $G_0(\delta)$, coincides with one of the solutions $x_0(t)$ on the first interval of continuity, except possibly, the δ_1-neighborhood of θ_1. Then on the interval $[\theta_1, \theta_2]$, it coincides with one of the solutions $x_1(t)$, except possibly, the δ_1-neighborhood of θ_1 and the δ_2-neighborhood of θ_2, etc. Finally, one can see that the integral curve of $\bar{x}(t)$ belongs to G^ϵ, it has exactly k meeting points with the surfaces Γ_i, θ_i^1, $i = 1, 2, \ldots, k$, $\left|\theta_i^1 - \theta_i\right| < \epsilon$ for all i and is continuable to $t = T$. The lemma is proved. □

Theorem 6.1.1. *Suppose conditions (N1)–(N6),(N11),(N12) are valid. Then the solution $x(t)$ of (6.1) has B-derivatives with respect to initial conditions, $(u_j(t), \nu_{ij}), j = 0, 1, 2, \ldots, n$, which satisfy the following variational equation:*

$$\begin{aligned} \frac{du}{dt} &= A(t)u, \\ \Delta u|_{t=\theta_i} &= P_i u, \\ \nu_i &= m_i u(\theta_i), \end{aligned} \tag{6.19}$$

with initial value $-f(t_0, x_0)$ if $j = 0$ and $e^j = \underbrace{(0, ..., 0, 1}_{j}, 0, ..., 0), j = 1, 2, \ldots, n$. In (6.19)

$$A(t) = f_x(t, x(t)), P_i = (f - f^+)\tau_{ix}(1 - \tau_{ix} f)^{-1} +$$

$$J_{ix}(\mathcal{I} + f\tau_{ix}(1 - \tau_{ix} f)^{-1})), m_i = (1 - \tau_{ix} f)^{-1}\tau_{ix}.$$

Proof. Consider differentiability in x_0. We use the last lemma and take $\delta > 0$ sufficiently small such that a solution $\bar{x}(t) = x(t, t_0, x_0 + \Delta x), ||\Delta x|| < \delta$, exists on the interval $[t_0, T]$, intersects each surface $\Gamma_i, i = 1, 2, \ldots, k$, once at moments $\theta_i^1, t_0 < \theta_1^1 < \ldots < \theta_k^1 < T$, which are near the moments of discontinuity of $x(t)$. One can choose small $\epsilon > 0$ such that (6.1) is B-equivalent to the system

$$\begin{aligned} \frac{dy}{dt} &= f(t, y), \\ \Delta y|_{t=\theta_i} &= W_i(y), \end{aligned} \tag{6.20}$$

in G^ϵ, where $W_i(x)$ are defined by using the map (6.7) in neighborhoods of points $x(\theta_i)$. Applying Lemma 6.1.2, one can find that $W_i(x)$ are continuously differentiable functions. Consequently, by Lemma 6.1.5, the solution $y(t, t_0, x_0)$ is continuously differentiable in initial data. The B-equivalence implies immediately that the corresponding solution $x(t, t_0, x_0)$ is also B-differentiable. We can easily find P_i and m_i, using equalities $P_i = W_{ix}(x(\theta_i))$ and formulas (6.8), (6.9). The theorem is proved. □

6.2 Differentiability with Respect to Parameters

Consider the following system of impulsive differential equations:

$$\begin{aligned} \frac{dx}{dt} &= f(t, x, \mu), \\ \Delta u|_{t=\tau_i(x,\mu)} &= J_i(x, \mu), \end{aligned} \tag{6.21}$$

where $(t, i, x) \in I \times \mathcal{A} \times G \subset \mathbb{R} \times \mathbb{Z} \times \mathbb{R}^n, \mu = (\mu_1, \ldots, \mu_m) \in G_\mu \subset \mathbb{R}^m$, where G_μ is an open set, m and n are fixed positive integers. Functions f, J_i, and τ_i are continuous and continuously differentiable in x and μ. We assume that conditions (N1)–(N6),(N11),(N12), accepted above for system (6.1), as well as inequalities $\tau_i(x,\mu) < \tau_{i+1}(x,\mu), i \in \mathcal{A}$, are valid for (6.21) uniformly with respect to μ in G_μ. Moreover, assume that $J_i, \tau_i \in C^{(r,r)}(G \times G_\mu)$, $f \in C^{(0,r,r)}(I \times G \times G_\mu) \cap C^{(r-1,r,r)}(N_h)$, where r is a positive integer, and N_h is a union of h-neighborhoods of surfaces $t = \tau(x,\mu), i \in \mathcal{A}$, in $I \times G \times G_\mu$.

Consider the system of ordinary differential equations

$$x' = f(t,x,\mu), \tag{6.22}$$

which is the differential part of (6.21), and maps $J(x,\mu) : G \times G_\mu \to \mathbb{R}^n$, $\tau(x,\mu) : G \times G_\mu \to \mathbb{R}$, which are continuously differentiable in x. Choose a point $(\kappa, x) \in I \times G$ and keep κ fixed. Let $t = \theta, \theta = \theta(x,\mu)$ be a meeting moment of $x(t) = x(t,\kappa,x,\mu)$, the solution of (6.22), with the surface $t = \tau(x,\mu)$. Moreover, assume that the solution $\bar{x}(t) = x(t,\theta,x(\theta) + I(x(\theta)),\mu)$ of the system exists on $\widehat{[\theta,\kappa]}$. Define the following B-map:

$$W(x,\mu) = \int_\kappa^\theta f(u,x(u),\mu)du + J(x + \int_\kappa^\theta f(u,x(u))du,\mu) + \int_\theta^\kappa f(u,\bar{x}(u),\mu)du. \tag{6.23}$$

Denote by $\tilde{G} \times \tilde{G}_\mu \subset G \times G_\mu$, the set of all points such that the map $W : x \to \bar{x}(\kappa)$ is defined. Similarly to Lemmas 6.1.3 and 6.1.4, the following assertion can be proved.

Lemma 6.2.1. $W(x,\mu), \theta(x,\mu) \in C^{(r,r)}(\tilde{G} \times \tilde{G}_\mu)$.

For a fixed $\mu_0 \in G_\mu$, there exists a unique solution $x(t,t_0,x_0,\mu_0)$, of (6.1) on some interval $[t_0,T], T > t_0$. Denote by $\theta_i, i = m, m+1,\ldots,k$ the moments of discontinuity of the solution. We assume that $t_0 \le \theta_m < \ldots < \theta_k < T$, $(t_0,x_0) \in G_m(\mu_0) = \{(t,x,\mu_0) : \tau_m(x,\mu_0) < t \le \tau_{m+1}(x,\mu_0), x \in G\}$. By introducing a new space variable μ, that is adding the equation $\mu' = 0$ to the system (6.21), we can find that the continuous dependence on the parameter is also valid.

Consider the disc $D((t_0,x_0),\delta) = \{(t_0,x_0,\mu) : \|\mu - \mu_0\| < \delta\}$ with the center (t_0,x_0,μ_0) and with the radius $\delta > 0$.

Denote by $x^j(t) = x(t,t_0,x_0,\mu_0 + \xi e_j), j = 1,2,\ldots,m$, a solution defined on $[t_0,T]$, and $\eta_i^j, i = 1,2,\ldots,k$, the moments of discontinuity of this solution.

Definition 6.2.1. The solution $x(t)$ is B-differentiable in $\mu_j, j = 1,2,\ldots,m$, if there exists $\delta > 0$ such that $(t_0,x_0,\mu_0 + \xi e_j) \in D((t_0,x_0),\delta) \cap G_m(\mu_0)$ implies that:

(1) there exist real constants $\beta_{ij}, i = 1,2,\ldots,k$, which satisfy

$$\theta_i - \eta_i^j = \beta_{ij}\xi + o(|\xi|); \tag{6.24}$$

(2) for all $t \notin \widehat{(\theta_i, \eta_i^j]}, i = 1, 2, \ldots, k$, it is true that

$$x^j(t) - x(t) = v_j(t)\xi + o(|\xi|), \tag{6.25}$$

where $v_j(t) \in \mathcal{PC}([t_0, T], \theta), \theta = \{\theta_i\}_{i=1,2,\ldots,k}$.

The pair $\{v_j(t), \{\beta_i^j\}\}$ is called a B-derivative of $x(t)$ with respect to μ_j.

Assume that the solution $x(t)$ meets each surface of discontinuity transversally, that is,

$$\tau_{ix}(x(\theta_i), \mu_0) f(\theta_i, x(\theta_i), \mu_0) \neq 1, \tag{6.26}$$

for all $i = m, m+1, \ldots, k$. Denote

$$A(t) = f_x(t, x(t), \mu_0),\ P_i = (f - f^+)\tau_{ix}(1 - \tau_{ix} f)^{-1} + J_{ix}(\mathcal{I} + f\tau_{ix}(1 - \tau_{ix} f)^{-1})),$$

$$g_j(t) = f_{\mu_j}(t, x(t), \mu_0),\ Q_j^i = J_{i\mu_j} + \tau_{i\mu_j}(f^+ - (\mathcal{I} + J_{ix})f)(1 - \tau_{ix} f)^{-1}, k_j^i = (1 - \tau_{ix} f)^{-1}\tau_{i\mu_j},$$

where values of functions are evaluated either at $(\theta_i, x(\theta_i), \mu_0)$ or $(\theta_i, x(\theta_i+), \mu_0)$, and the last case is indicated with the upper index $+$.

Similarly to Theorem 6.1.1, by using the map $W(x, \mu)$, we can show that the following assertion is valid.

Theorem 6.2.1. *Suppose that conditions (N1)–(N6),(N11),(N12), accepted above for system (6.1) are valid for (6.21) uniformly with respect to* μ *in* G_μ. *Moreover, there exists a number* $\kappa > 0$ *such that* $\tau_{m+1}(x_0, \mu) \geq \tau_{m+1}(x_0, \mu_0)$ *for all* $\|\mu - \mu_0\| < \kappa$.

Then the solution $x(t)$ *has the* B*-derivative with respect to* $\mu_j, j = 1, 2, \ldots, m$, *which satisfies the following variational system*

$$\begin{aligned}
\frac{dv}{dt} &= A(t)v + g_j(t),\\
\Delta v|_{t=\theta_i} &= P_i v + Q_j^i,\\
\beta_i &= k_j^i v(\theta_i),
\end{aligned} \tag{6.27}$$

with $v_j(t_0) = 0$.

6.3 Higher Order B-Derivatives

In this section, we assume that surfaces $\Gamma_i, i \in \mathcal{A}$, are placed in $I \times G$ with their h-neighborhoods for some fixed positive number h. Let us denote the union of the neighborhoods by N_h. We assume that $J_i, \tau_i \in C^{(r)}(G), f \in C^{(0,r)}$

$(I \times G) \cap C^{(r-1,r)}(N_h)$, where r is a positive integer. That is, functions $J_i, \tau_i, i \in \mathcal{A}$, are r times continuously differentiable in G. The function f is continuous in t, and r times continuously differentiable in $x \in G$, and additionally it is $r-1$ times continuously differentiable in t if the point (t, x) is in N_h. Consider the solution $x(t) = x(t, t_0, x_0)$ of system (5.45) again. In what follows, assume that the solution $x(t)$ meets each surface of discontinuity transversally, that is, (6.26) is valid. Denote by $\bar{x}(t) = x(t, t_0 + \xi_0, x_0 + \Delta x), \Delta x = (\xi_1, \xi_2, \ldots, \xi_n), \xi = (\xi_0, \xi_1, \xi_2, \ldots, \xi_n)$ another solution of (6.1). Denote by $\eta_i, i \in \mathcal{A}$, the moments of discontinuity of this solution, and $B((t_0, x_0), \delta)$ the ball in $\mathbb{R} \times \mathbb{R}^n$ with the center at (t_0, x_0) and the radius δ.

Definition 6.3.1. The solution $x(t)$ has B-derivatives of up to r-th order, inclusive, with respect to t_0 and $x_0^j, j = 1, 2, \ldots, n$, if there exist functions $u_{1i}(t), u_{2ij}(t), \ldots, u_{rij\ldots s}(t) \in \mathcal{PC}^1([t_0, T], \theta), \theta = \{\theta_i\}_{i=1,2,\ldots,k}$ and constants $\nu_{1i}^l, \nu_{2ij}^l, \ldots, \nu_{rij\ldots s}^l, l = 1, 2, \ldots, k$, which are symmetric with respect to permutation of indices, such that if $(t_0 + \xi, x_0 + \Delta x) \in B((t_0, x_0)\delta) \cap G_m$, for sufficiently small $\delta > 0$, then:

(1)

$$\theta_l - \eta_l = \sum_{i=0}^{n} \nu_{1i}^l \xi_i + \ldots + \sum_{i,j,\ldots s=0}^{n} \nu_{rij\ldots s}^l \xi_i \xi_j \ldots \xi_s + o(||\xi||^r), \tag{6.28}$$

where $l = 1, 2, \ldots, k$;

(2) for all $t \notin \widehat{(\theta_i, \eta_i]}, i = 1, 2, \ldots, k$, it is true that

$$\bar{x}(t) - x(t) = \sum_{i=0}^{n} u_{1i}(t)\xi_i + \ldots + \sum_{i,j,\ldots s=0}^{n} u_{rij\ldots s}(t)\xi_i \xi_j \ldots \xi_s + o(||\xi||^r). \tag{6.29}$$

The pairs $\{u_{1i}(t), \{\nu_{1i}^l\}\}, \ldots, \{u_{rij\ldots s}(t), \{\nu_{rij\ldots s}^l\}\}$ are called the B-derivatives of the solution $x(t)$ with respect to initial conditions.

Lemma 6.3.1. *Suppose $f \in C^{(0,r)}(I \times G), W_i \in C^{(r)}(G)$. Then the solution $y(t, t_0, x_0)$ of system (6.15) has B-derivatives of up to r-th order, inclusive, with respect to $t_0, x_0^j, j = 1, 2, \ldots, n$, on the interval $[t_0, T]$. The formulas (6.29) are valid for all $t \in [t_0, T]$. That is, constants ν in (6.28) are all equal to zero.*

Proof. By Lemma 6.1.5, solution $y(t, t_0, x_0)$ has B-derivatives of the first order, which are solutions of the variational system (6.16). The right-hand side of the system satisfies the conditions of Lemma 6.1.5. Hence, there exist B-derivatives for $y(t, t_0, x_0)$ of the second order, $(\nu = 0)$. Repeat similar discussion $r-1$ times to obtain the complete proof of the lemma.

Theorem 6.3.1. *Suppose conditions (N1)–(N6), (N12) are valid, and* $J_i, \tau_i \in C^{(r)}(G)$, $f \in C^{(0,r)}(I \times G) \cap C^{(r-1,r)}(N_h)$, *where* r *is a positive integer. Then the solution* $x(t)$ *of (6.1) has* B*-derivatives of up to* r*-th order, inclusive, with respect to* t_0, x_0^j, $j = 1, 2, \ldots, n$.

Proof. Let us consider (6.1) in the domain G^h, where systems (6.1) and (6.15) are B-equivalent. One can show that the map $W_i, i = 1, 2, \ldots, k$, is defined in a neighborhood G_i of the point $x(\theta_i)$. Without loss of generality we assume that $G_i = G$ for all i. By Lemma 6.1.4, one has $W_i \in C^{(r)}(G), i = 1, 2, \ldots, k$. Consequently, by the last lemma there exist B-derivatives of $y(t, t_0, x_0)$ to r-th order, inclusive. It is obvious that the first components, u, of the B-derivatives is the first component of the B-derivatives of $x(t, t_0, x_0)$. By using (6.10), one can verify (6.28). The theorem is proved. □

Consider the solutions $x(t) = x(t, t_0, x_0, \mu_0)$ and $\bar{x}(t) = x(t, t_0, x_0, \mu_0 + \Delta\mu)$ of system (6.21), again. Denote by $\eta_i, i = 1, 2, \ldots, k$, the moments of discontinuity of $\bar{x}(t)$.

Definition 6.3.2. The solution $x(t)$ has B-derivatives of up to r-th order, inclusive, with respect to parameters $\mu_j, j = 1, 2, \ldots, m$, on $[t_0, T]$ if there exist functions $v_{1i}(t), v_{2ij}(t), \ldots, v_{rij\ldots s}(t) \in \mathcal{PC}^1([t_0, T], \theta), \theta = \{\theta_i\}_{i=1,2,\ldots,k}$ and constants $\beta_{1i}^l, \beta_{2ij}^l, \ldots, \beta_{rij\ldots s}^l, l = 1, 2, \ldots, k$, which are symmetric with respect to permutation of indices, such that if $(t_0, x_0, \mu) \in D((t_0, x_0), \delta) \cap G_m(\mu_0)$ for some $\delta > 0$, then:

(1)

$$\theta_l - \eta_l = \sum_{i=1}^{n} \beta_{1i}^l \xi_i + \ldots + \sum_{i,j,\ldots s=1}^{n} \beta_{rij\ldots s}^l \xi_i \xi_j \ldots \xi_s + o(\|\xi\|^r), \quad (6.30)$$

where $l = 1, 2, \ldots, k$;

(2) for all $t \notin \widehat{(\theta_i, \eta_i]}, i = 1, 2, \ldots, k$, it is true that

$$\bar{x}(t) - x(t) = \sum_{i=1}^{n} v_{1i}(t)\xi_i + \ldots + \sum_{i,j,\ldots s=1}^{n} v_{rij\ldots s}(t)\xi_i \xi_j \ldots \xi_s + o(\|\xi\|^r). \quad (6.31)$$

The pairs $\{v_{1i}(t), \{\beta_{1i}^l\}\}, \ldots, \{v_{rij\ldots s}(t), \{\beta_{rij\ldots s}^l\}\}$ are called B-derivatives of the solution $x(t)$ in $\mu_j, j = 1, 2, \ldots, m$.

Using Lemma 6.2.1, similar to the proof of Theorem 6.3.1, one can prove that the following theorem is valid.

Theorem 6.3.2. *Suppose conditions (N1)–(N6), (N12) are valid uniformly with respect to $\mu \in G_\mu$, and $J_i, \tau_i \in C^{(r,r)}(G \times G_\mu)$, $f \in C^{(0,r,r)}(I \times G \times G_\mu) \cap C^{(r-1,r,r)}(N_h)$, where r is a positive integer. Then the solution $x(t)$ of (6.21) has B-derivatives of up to r-th order, inclusive, with respect to μ_j, $j = 1,2,\ldots,m$, if there exists a number $\kappa > 0$ such that $\tau_{m+1}(x_0,\mu) \geq \tau_{m+1}(x_0,\mu_0)$ for all $\|\mu - \mu_0\| < \kappa$.*

6.4 B-Analyticity Property

In this part of the book, assume that in addition to conditions (N1)–(N6), (N12), which are valid for all $\mu \in G_\mu$, functions J_i, τ_i, and f are holomorphic in x and μ on $G \times G_\mu$, and f is holomorphic in t, x, and μ in the region N_h. Let us choose a point $(\kappa, x_0, 0)$ in $I \times G \times G_\mu$, and fix κ. Denote by $t = \theta$ a meeting moment of a solution $x(t,\kappa,x_0,0)$ of (6.6) and the surface $t = \tau(x,0)$ transversally. Assume that there exists a number $\kappa > 0$ such that $\tau(x_0,\mu) \geq \tau(x_0,0)$ for all $\|\mu\| < \kappa$. Moreover, f is holomorphic in t in a neighborhood of $(\theta, x(\theta+), 0)$. Consider a solution $\bar{x}(t) = x(t,\kappa,x,\mu)$ of (6.6). If $||\mu||$ and $||x - x_0||$ are sufficiently small, then $\bar{x}(t)$ intersects the surface Γ transversally at moment $\bar{\theta} = \theta(x,\mu)$ near to θ. We assume also that (6.6) admits a solution $x_1(t) = x(t,\bar{\theta},\bar{x}(\bar{\theta}) + I(\bar{x}(\bar{\theta}),\mu)$, defined on $\widehat{[\kappa,\theta]}$, and introduce the map $W : (x,\mu) \to x_1(\kappa) - x$.

Lemma 6.4.1. *$\theta(x,\mu)$ is holomorphic at $(x_0, 0)$.*

Proof. Apply Poincaré's expansion theorem [98] to show that if t is near $t = \theta$ then the expansion

$$x(t) = \sum C_{p\alpha\ldots\lambda a\ldots l}(t-\theta)^p (x - x_1^0)^\alpha \ldots (x - x_n^0)^\lambda \mu_1^a \ldots \mu_m^l \quad (6.32)$$

is valid. Since the solution meets the surface of discontinuity transversally, if $||x - x_0||$ and $||\mu||$ are sufficiently small there exists a unique solution of the equation $\theta = \tau(\bar{x}(\theta),\mu)$. The function $\bar{\theta} = \theta(x,\mu)$ is defined as an implicit function, from the equation $\Psi(\bar{\theta},\mu) \equiv \bar{\theta} - \tau(\bar{x}(\bar{\theta}),\mu) = 0$, where $\Psi_\theta(\theta,0) \neq 0$. Using the theorem on the holomorphic implicit function [98], one can find

$$\bar{\theta} = \sum B_{p\alpha\ldots\lambda a\ldots l}(t-\theta)^p (x - x_1^0)^\alpha \ldots (x - x_n^0)^\lambda \mu_1^a \ldots \mu_m^l$$

such that $\bar{\theta}(x_0,0) = \theta$. The lemma is proved. □

By using $W(x,\mu) = x(\kappa,\bar{\theta},x(\bar{\theta},x,\mu) + I(x(\bar{\theta},x,\mu),\mu)) - x$, Lemma 6.4.1, and the theorem on substitution of a series into a series [59] one can prove that the following assertion is valid.

Lemma 6.4.2. *$W(x,\mu)$ is holomorphic at $(x_0,0)$.*

Let us consider the main subject of our discussion. Denote by $x(t) = x(t,t_0,x_0,0)$ a solution of (6.21) with $\mu = 0$, and assume that it exists on an interval $[t_0,T], T > t_0$, with points of discontinuity $t_0 < \theta_1 < \ldots < \theta_k < T$. From the previous discussion, it implies that a solution $\bar{x}(t) = x(t,t_0,x,\mu)$ with sufficiently small $||x - x_0||$ and $||\mu||$ exists on the interval $[t_0,T]$. Denote by $\eta_i, i = 1,2,\ldots,k$, the points of discontinuity of this solution. One can find a neighborhood of the integral curve of $x(t)$ on $[t_0,T]$ such that system (6.21) and

$$\begin{aligned} \frac{dy}{dt} &= f(t,y,\mu),\\ \Delta y|_{t=\theta_i} &= W_i(y,\mu), \end{aligned} \tag{6.33}$$

are B-equivalent there. Since functions W_i are specified W for $\kappa = \theta_i$, by Lemma 6.4.2, they are analytic at points $(\theta_i, x(\theta_i),0)$. There exists a solution $\phi(t) = y(t,t_0,x_0,0)$ of system

$$\begin{aligned} \frac{dy}{dt} &= f(t,y,0),\\ \Delta y|_{t=\theta_i} &= W_i(y,0), \end{aligned} \tag{6.34}$$

such that $\phi(t) = x(t)$ for all $t \in [t_0,T]$.

Definition 6.4.1. The solution $y(t) = y(t,t_0,x,\mu)$ of (6.33) is expanded in power series of coordinates $x - x_0$ and μ if

$$y(t) = \sum A_{\alpha\ldots\lambda a\ldots l}(t)(x - x_1^0)^{\alpha}\ldots(x - x_n^0)^{\lambda}\mu_1^a\ldots\mu_m^l, t \in [t_0,T],$$

where coefficients $A \in \mathcal{PC}_r([t_0,T],\theta)$.

Lemma 6.4.3. *The solution $y(t) = y(t,t_0,x,\mu)$ of system (6.33) is expanded in power series of coordinates of $x - x_0$ and μ.*

Proof. By continuity, if $||x-x_0||$ and $||\mu||$ are sufficiently small, then solution $y(t)$ exists on $[t_0,T]$. Let us show that

$$y(t) = \sum A_{\alpha\ldots\lambda a\ldots l}(t)(x - x_1^0)^{\alpha}\ldots(x - x_n^0)^{\lambda}\mu_1^a\ldots\mu_m^l,$$

where coefficients A are piecewise continuous functions with discontinuities at points θ_i. Indeed, apply the Poincaré' expansion theorem to obtain a series on the interval $[t_0,\theta_1]$. Now, using analyticity of J_1 one can find that the value $y(\theta_i+) = y(\theta_i) + W_1(y(\theta_1),\mu)$ has also a series expansion. Consequently, considering the solution of (6.6) with initial condition $(\theta_i, x(\theta_i))$ on $[\theta_1,\theta_2]$ and using the lemma on substitution of a series into a series, we find that the expansion is valid on $(\theta_1,\theta_2]$. Proceeding for all i, one comes to the complete proof of the assertion. The lemma is proved. □

Definition 6.4.2. The solution $\bar{x}(t) = x(t, t_0, x, \mu)$ of (6.21), with discontinuity moments $\eta_i, i = 1, 2, \ldots, k$, is expanded in power series of coordinates $x - x_0$ and μ (or it B-analytically depends on x and μ) if for all $t \notin \widehat{(\theta_i, \eta_i]}$ it is true that

$$\bar{x}(t) = \sum A_{\alpha\ldots\lambda a\ldots l}(t)(x - x_1^0)^{\alpha} \ldots (x - x_n^0)^{\lambda} \mu_1^a \ldots \mu_m^l,$$

and

$$\theta_i - \eta_i = \sum D^i_{\alpha\ldots\lambda a\ldots l}(x - x_1^0)^{\alpha} \ldots (x - x_n^0)^{\lambda} \mu_1^a \ldots \mu_m^l,$$

where coefficients $A \in \mathcal{PC}_r([t_0, T], \theta)$ and D are real numbers.

From the last lemma and the B-equivalence it follows that the next theorem is valid.

Theorem 6.4.1. *The solution $\bar{x}(t) = x(t, t_0, x, \mu)$ of system (6.21) is expanded in power series of coordinates $x - x_0$ and μ.*

The following assertion is an easy corollary of the last theorem.

Theorem 6.4.2. *For each fixed $\bar{t} \in [t_0, T]$ the function $x(\bar{t}, t_0, x, \mu)$ is an analytic function of x and μ in a neighborhood of the point* $(x_0, 0)$.

6.5 B-Asymptotic Approximation of Solutions

In this part of the chapter, we investigate the problem of asymptotic approximation with respect to the small parameter of solutions of impulsive differential equations with impulses on surfaces. The results obtained here are development of previous parts of the present chapter. Consider system (6.21), assuming this time that the dimension of the parameter space μ is one, $m = 1$. We assume that for all $\mu \in (-\mu_0, \mu_0)$, where μ_0 is a fixed positive number, conditions (N1)–(N6),(N11),(N12), are valid, and the following higher order differentiability is fulfilled, $J_i, \tau_i \in C^{(r,r)}(G \times G_\mu)$, $f \in C^{(0,r,r)}(I \times G \times G_\mu) \cap C^{(r-1,r,r)}(N_h)$, where r is a positive integer, and N_h is a union of h-neighborhoods of surfaces $t = \tau(x, \mu), i \in \mathcal{A}$, in $I \times G \times G_\mu$.

Denote by $x_0(t) = x(t, t_0, x_0, 0)$ a solution of (6.21), where $\mu = 0$, that is, $x_0(t)$ is a solution of the system

$$\begin{aligned} \frac{dx}{dt} &= f(t, x, 0), \\ \Delta x|_{t=\tau_i(x,0)} &= J_i(x, 0), \end{aligned} \tag{6.35}$$

and assume that it exists on an interval $[t_0, T], T > t_0$, with points of discontinuity $t = \theta_i, t_0 < \theta_1 < \ldots < \theta_k < T$.

From the earlier discussion, it implies that a solution $x(t, \mu) = x(t, t_0, x_0, \mu)$ with sufficiently small $|\mu|$ exists on the interval $[t_0, T]$. Denote by $\eta_i, i = 1, 2, \ldots, k$, the points of discontinuity of this solution.

We say that the solution $x(t,\mu)$ has a B-asymptotic approximation if for sufficiently small $|\mu|$ and for all $t \in [t_0, T]$ outside the intervals $\widehat{[\theta_i, \eta_i]}, i = 1, 2, \ldots, k$, the following equality is valid,

$$x(t,\mu) = \sum_{j=1}^{r} x_j(t)\mu^j + O(\mu^{r+1}), \tag{6.36}$$

where $x_j(t)$ are piecewise continuous vector-valued functions from $\mathcal{PC}([t_0, T], \theta)$, $\theta = \{\theta_i\}_{i=1,2,\ldots,k}$. Moreover, for all $i = 1, 2, \ldots, k$,

$$\eta_i - \theta_i = \sum_{j=1}^{r} \kappa_{ij}\mu^j + O(\mu^{r+1}), \tag{6.37}$$

where κ_{ij} are real constants.

By the B-equivalence, $x_0(t)$ is a solution of (6.35) and (6.34). Moreover, one can find a sufficiently small neighborhood of the integral curve of $x_0(t)$ on $[t_0, T]$, where systems (6.21) and (6.33) are B-equivalent. The existence of the expansions (6.36) and (6.37) is proved in Theorem 6.3.2. We consider here the problem of determining the coefficients $x_j(t)$ and κ_{ij}. By virtue of the correspondence established between solutions $x(t,\mu)$ and $y(t,\mu)$ of systems (6.21) and (6.33) above, it suffices to determine the x_j beginning with (6.33), taking into account the fact that the asymptotic formula

$$y(t,\mu) = \sum_{j=1}^{r} x_j(t)\mu^j + O(\mu^{r+1}), \tag{6.38}$$

holds for all points in $[t_0, T]$.

Substituting the last expression into (6.33) and using the smoothness of f and W_i, we find that for each $j = 1, 2, \ldots, r$, the coefficient $x_j(t)$ is a solution of the Cauchy problem $x_j(t_0) = 0$ for the equation

$$\begin{aligned} \frac{dx}{dt} &= \frac{f(t, x_0(t), 0)}{dx}x + F(t, x_0, x_1, \ldots, x_{j-1}), \\ \Delta x|_{t=\theta_i} &= \frac{W_i(x, 0)}{dx}x + G_i(x_0, x_1, \ldots, x_{j-1}), \end{aligned} \tag{6.39}$$

where functions F and G_i are completely determined by $x_0, x_1, \ldots, x_{j-1}$ and the partial derivatives of f and V_i of order up and including j, evaluated for $x = x_0(t)$ and $\mu = 0$.

Let us determine the partial derivatives of the W_i at the points $(x_0(\theta_i), 0)$. Fix i and for brevity we omit the index i in what follows. We let $\theta_i = \theta$, and if $x = x_0(\theta), t = \theta$, and $\mu = 0$, or if $x = x_0(\theta+), t = \theta$, and $\mu = 0$, then all functions used below will be written without showing the values of the arguments, and we will distinguish between these two cases by using a $+$ superscript in the second case.

Also, we use the subscripts x, t, and μ to indicate partial derivatives. We will assume that x, f, J, and W and their derivatives are column vectors, and that derivatives of τ and θ are row vectors. We define the product of vectors and matrices according to the usual rule of multiplying rectangular matrices.

The moment of discontinuity $t = \eta$ of the solution $x(t, \mu)$ is determined from the equation $t = \tau(x(t, \mu), \mu)$ as a function $\eta = \eta(x, \mu)$, where $x = x(\eta, \mu)$. Therefore, applying the known implicit function theorem and passing to the limit as $\mu \to 0$, we get

$$\begin{aligned}
&\eta_x = \tau_x(1 - \tau_x f)^{-1}, \eta_\mu = \tau_x(1 - \tau_x f)^{-1}\tau_\mu,\\
&\eta_{x\mu} = \tau_{x\mu}(1 - \tau_x f)^{-1} + \tau_x(\tau_{x\mu} f + \tau_x f_\mu)(1 - \tau_x f)^{-2},\\
&\eta_{\mu\mu} = \tau_\mu(1 - \tau_x f)^{-1} + \tau_\mu(\tau_{x\mu} f + \tau_x f_\mu),\\
&\eta_{xx_j} = 2\tau_{xx_j}(1 - \tau_x f)^{-1} + \tau_x(2\tau_{xx_j} f + \tau_x f_{x_j})(1 - \tau_x f)^{-2}. \qquad (6.40)
\end{aligned}$$

Applying the resulting expressions in a similar way, starting from (6.23), we find that

$$\begin{aligned}
&W_x = \eta_x(f - f^+) + J_x(\mathcal{I} + \eta_x f), W_\mu = (f - f^+)\eta_\mu + J_x f \eta_\mu + J_\mu,\\
&W_{x\mu} = \eta_x(f_t - f_t^+)\eta_\mu + \eta_{x\mu}(f - f^+) + (f_x - f_x^+(\mathcal{I} + J_x))\eta_\mu +\\
&(\sum_{k=1}^{n} J_{xx_k} f_k \eta_\mu + J_{x\mu})(\mathcal{I} + \eta_x f) + J_x(\eta_x f_\mu + f_x\eta_\mu + \eta_{x\mu} f),\\
&W_{xx_j} = \eta_x((f_t - f_t^+)\eta_{x_j} + f_{x_j} - f_{x_j}^+) + \eta_{xx_j}(f - f^+) +\\
&\eta_{x_j}(f_x - f_x^+ - f_x^+ J_x) + \sum_{k=1}^{n} J_{xx_k}(\delta_{kj} + f_k\eta_{x_j})(\mathcal{I} + \eta_x f) +\\
&J_x(\eta_x(f_{x_j} + f_t\eta_{x_j}) + \eta_{xx_j} f + \eta_{x_j} f_x),\\
&W_{\mu\mu} = (f_t - f_t^+)\eta_\mu^2 + (f - f^+)\eta_{\mu\mu} + (f_\mu + J_{x\mu} f)\eta_\mu +\\
&\sum_{k=1}^{n} J_{xx_k} f_k \eta_\mu(f\eta_\mu + J_\mu) + J_x(f_t\eta_\mu^2 + 2f_\mu\eta_\mu + f\eta_{\mu\mu}) +\\
&\sum_{k=1}^{n} J_{\mu x_k} f_k + J_{\mu\mu}. \qquad (6.41)
\end{aligned}$$

where δ_{ij} is the Kronecker symbol. It is clear that in this way it is possible to calculate the derivatives of W up to and including order r at the point $(x(\theta_i), 0)$ from the values of the derivatives of f, J, and τ at points $(\theta, x_0(\theta), 0)$ and $(\theta, x_0(\theta+), 0)$. In addition, the coefficients in (6.37) can be determined by starting from (6.40).

Notes

Some results on differentiable dependence of first order for solutions of impulsive systems on initial conditions and parameters can be found in [45, 64, 74, 141, 142]. All definitions of differentiability of solutions, higher order differentiable dependence, and analyticity of solutions of differential equations with nonfixed moments of impulses were given in papers [24, 25, 32, 34, 35] for the first time, as well as corresponding theorems and their proofs. Further applications and development of the results for autonomous equations and Filippov's differential equations were made in [1, 3, 4, 14, 27, 36]. The results of the chapter show that the method of B-equivalence is suitable to realize any type of smoothness for differential equations with variable moments of discontinuities. The variational equations are presented for the first time with the second component of the B-derivatives.

Chapter 7
Periodic Solutions of Nonlinear Systems

In this part of the book, we investigate, by applying methods developed in the previous chapters, existence and stability of periodic solutions of quasilinear systems with variable moments of impulses.

7.1 Quasilinear Systems: the Noncritical Case

Consider the following (ω, p)-periodic system of differential equations with impulses on surfaces

$$x' = A(t)x + f(t) + \mu\phi(t, x, \mu),$$
$$\Delta x|_{t=\theta_i+\mu\tau_i(x,\mu)} = B_i x + I_i + \mu\Psi_i(x, \mu), \tag{7.1}$$

where $(t, x) \in \mathbb{R} \times \mathbb{R}^n, \mu \in (-\mu_0, \mu_0)$, μ_0 is a fixed positive number, real valued elements of $n \times n$ matrix $A(t)$ are continuous functions, and constant real valued $n \times n$ matrices $B_i, i \in \mathbb{Z}$, satisfy condition (4.2). Coordinates of $f(t) : \mathbb{R} \to \mathbb{R}^n$ are continuous functions, and $J_i, i \in \mathbb{Z}$, is a sequence of vectors from $\mathbb{R}^n$. Being an (ω, p)-periodic system means that there exist a positive real number ω and a positive integer p such that $A(t+\omega) = A(t), f(t+\omega) = f(t), \phi(t+\omega, x, \mu) = \phi(t, x, \mu)$, for all $t \in \mathbb{R}, x \in \mathbb{R}^n, \mu \in (-\mu_0, \mu_0)$, and $B_{i+p} = B_i, I_{i+p} = I_i, \Psi_{i+p}(x, \mu) = \Psi_i(x, \mu), \theta_{i+p} = \theta_i + \omega, \tau_{i+p}(x, \mu) = \tau_i(x, \mu)$, for all $i \in \mathbb{Z}, x \in \mathbb{R}^n$, $\mu \in (-\mu_0, \mu_0)$. We assume also that $\phi \in C^{(0,1,1)}(\mathbb{R} \times \mathbb{R}^n \times (-\mu_0, \mu_0)), \tau_i$, $\Psi_i \in C^{(1,1)}(\mathbb{R}^n \times (-\mu_0, \mu_0)), i \in \mathbb{Z}$.

It is obvious that θ is a B-sequence. That is, the infinite sequence satisfies $|\theta_i| \to \infty$ as $|i| \to \infty$. Beside the system (7.1), consider the nonperturbed system

$$x' = A(t)x + f(t),$$
$$\Delta x|_{t=\theta_i} = B_i x + I_i. \tag{7.2}$$

M. Akhmet, *Principles of Discontinuous Dynamical Systems*,
DOI 10.1007/978-1-4419-6581-3_7, © Springer Science+Business Media, LLC 2010

Theorem 7.1.1. *Assume that (7.2) has a unique nontrivial ω-periodic solution $x_0(t)$. If $|\mu|$ is sufficiently small, then (7.1) admits an ω-periodic solution that converges in the B-topology to $x_0(t)$ as $\mu \to 0$.*

Proof. Without loss of generality, we assume that the periodic solution $x_0(t) = x(t, 0, x_0, 0)$ has moments of discontinuity θ_i such that $0 < \theta_1 < \ldots < \theta_p < \omega$. Let $x(t) = x(t, 0, \xi, \mu)$ be a solution of (7.1), $x(0) = \xi$. If $|\mu|$ is sufficiently small, then from results of the last chapter it follows that $x(t)$ has moments of discontinuity η_i such that $0 < \eta_1 < \ldots < \eta_p < \omega$, and there is a neighborhood of the point $(0, x_0)$, which does not intersect all surfaces of discontinuity $t = \theta_i + \mu\tau(x, \mu)$. By the Poincaré criterion, the solution $x(t)$ is a periodic function if and only if the equation

$$\Phi(y, \mu) \equiv x(\omega, 0, y, \mu) - y = 0 \tag{7.3}$$

is satisfied with $y = \xi$. It follows, from the implicit function theorem that (7.3) has a solution if $\det \Phi_y(x_0, 0) \neq 0$, and the matrix is continuous in a neighborhood of $(x_0, 0)$. By Theorem 6.1.1, B-derivatives of $x(t)$ in ξ_j, $j = 1, 2, \ldots, n$, form a fundamental matrix $Z(t, y, \mu)$, $Z(0, y, \mu) = \mathcal{I}$, of the variational system. If $\mu = 0$, then the variational system has the form

$$\begin{aligned} x' &= A(t)x, \\ \Delta x|_{t=\theta_i} &= B_i x. \end{aligned} \tag{7.4}$$

The uniqueness of the periodic solution $x_0(t)$ implies that

$$\det \Phi_y(x_0, 0) = \det(Z(\omega, y, 0) - \mathcal{I}) \neq 0.$$

Hence, (7.3), for sufficiently small $|\mu|$, has a unique solution in a small neighborhood of x_0, and this solution is as much near x_0 as μ is close to 0. By Lemma 6.1.6, we have that the ω-periodic solution $x(t)$ converges to $x_0(t)$ as $\mu \to 0$. The theorem is proved. □

Fix positive h such that N_h, the union of h-neighborhoods of points $\theta_i, i = 1, 2, \ldots, p$, is placed in $[0, \omega]$. Similarly to the proof of the last theorem, one can verify that the following assertion is valid.

Theorem 7.1.2. *Assume that (7.2) has a unique nontrivial ω-periodic solution $x_0(t)$, functions ϕ, Ψ, τ_i are analytic in $x \in G$, and, moreover, functions A, f, ϕ are analytic in $t \in N_h$. Then for a sufficiently small $|\mu|$ system (7.1) admits an ω-periodic solution that converges in the B-topology to $x_0(t)$, as $\mu \to 0$. The solution is analytic at $\mu = 0$.*

Let us formulate the following assertion without proving it.

Theorem 7.1.3. *Suppose all multipliers of (7.4) are inside the unit disc. If $|\mu|$ is sufficiently small, then the ω-periodic solution of (7.1) is B-asymptotically stable.*

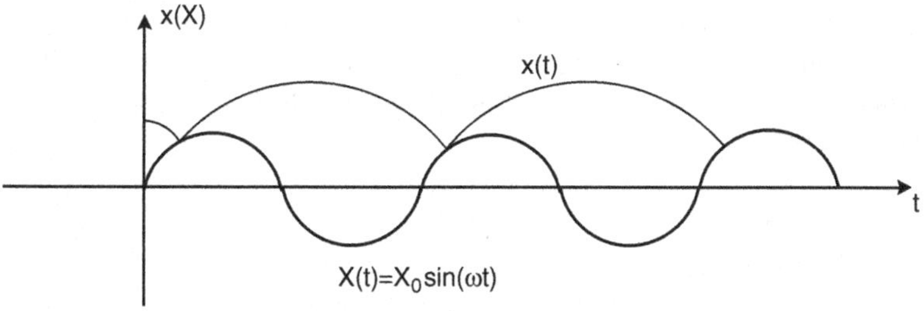

Fig. 7.1 The first coordinate, $x(t)$, of the periodic solution $\psi(t)$

Example 7.1.1. Consider system (1.3), which is constructed in Example 1.1 as a model of a mechanical system consisting of a bead bouncing on a table. In [89], it is proven that the mechanical model considered with the condition

$$\omega^2 \geq \frac{\pi g}{X_0}\frac{1-R}{1+R} \tag{7.5}$$

admits a periodic discontinuous motion $\psi(t)$ with period $T = \frac{2\pi}{\omega}$ (see Fig. 7.1).

The solution has the initial value $x_0 = X_0\sqrt{1-\cos^2(\phi)}$, $x_0' = \frac{\pi g}{X_0 \omega}$ at the moment of the impact $t = \phi$, which is defined by the relation

$$\cos(\phi) = \frac{\pi g}{X_0\omega^2}\frac{1-R}{1+R}. \tag{7.6}$$

Applying the Poincaré map, it was shown that the periodic motion is asymptotically stable if

$$\frac{\pi g}{X_0}\frac{1-R}{1+R} < \omega^2 < \{[\frac{\pi g}{X_0}\frac{1-R}{1+R}]^2 + [\frac{2g(1+R^2)}{X_0(1+R)^2}]^2\}^{\frac{1}{2}}. \tag{7.7}$$

It is obvious that (1.3) is a highly idealized model. It takes no account of drag in the medium, the unavoidable perturbations of the table, and possible elastic couplings. Hence, it is natural to consider a system of the form

$$\begin{aligned}
x_1' &= x_2,\\
x_2' &= -g + \mu f(t,x,\mu),\\
\Delta x_2|_{t=\tau_i(x_1)+\mu\kappa_i(x,\mu)} &= (1+R)[x_0\omega\cos(\omega\tau_i(x_1)) - x_2] + \mu I_i(x,\mu),
\end{aligned} \tag{7.8}$$

where μ is a small parameter.

We assume that functions f, I, and κ have continuous second-order partial derivatives with respect to x_1, x_2, and μ, and the function f is continuously differentiable with respect to t. The variational system around the periodic solution $\psi(t)$ has the form

$$\begin{aligned} u_1' &= u_2, \\ u_2' &= 0, \\ \Delta u_2|_{t=\phi} &= \frac{\omega}{2}[\frac{(1+R)^2}{\pi} - b(1-R)^2)]u_1 - (1+R)u_2, \end{aligned} \qquad (7.9)$$

where $b = \sqrt{\cos^{-2}(\phi) - 1}$. The characteristic equation for (7.9) is

$$\rho^2 + (\pi b(1-R^2) - (1+R^2))\rho + R^2 = 0.$$

We have from the last equation that (7.9) does not have a unique multiplier provided

$$R = 1 \quad \text{or} \quad \omega^2 \neq \frac{\pi g}{X_0}\frac{1-R}{1+R}.$$

A necessary and sufficient condition for the multipliers to be situated inside the unit circle is the inequality (7.7).

Assume that $R \neq 1$, and suppose that the function f is periodic with period $T = \frac{2\pi}{\omega}$ with respect to t, and for all $x, \mu, i \in \mathbb{Z}$, the equalities $I_{i+1} = I_i, \kappa_{i+1} = \kappa_i$ are satisfied. Then one can see that the system is $(T, 1)$-periodic, and from relations (7.5) and (7.7), we find, according to Theorems 7.1.1 and 7.1.3, that for a sufficiently small μ system (7.8) admits a unique T-periodic B-asymptotically stable solution, which converges to $\psi(t)$, as $\mu \to 0$.

In the sequel, we proceed with the investigation of system (5.38) and find the conditions for the existence of a unique bounded solution, a periodic solution, and asymptotical stability of these solutions. In this part of the section, we want to see the use of the reduction method for this system. Beside it, we consider the following equations with impulse actions at fixed moments

$$\begin{aligned} \frac{dx}{dt} &= A(t)x + f(t,x), \\ \Delta x|_{t=\theta_i} &= B_i x + I_i(x), \end{aligned} \qquad (7.10)$$

where $t \in \mathbb{R}, x \in \mathbb{R}^n, \theta = \{\theta_i\}, i \in \mathbb{Z}$, is a B-sequence, entries of $n \times n$ matrix $A(t)$ are continuous real valued functions, $B_i, i \in \mathbb{Z}$, are constant real valued square matrices of order n. That is, system (7.10) differs from (5.38) only by absence of perturbations $\tau_i(x)$.

For the convenience of discussion we will reformulate conditions $(Q1)$–$(Q5)$ of Sect. 5.8.

So, we assume that for system (7.10) the following conditions are valid:

(P1) there exists a positive constant κ such that $\theta_{i+1} - \theta_i \geq \kappa, i \in \mathbb{Z}$;

(P2) there exists a positive constant l such that for all $t \in \mathbb{R}, x, y \in \mathbb{R}^n, i \in \mathbb{Z}$, the following inequality is valid

$$\|f(t,x) - f(t,y)\| + \|I_i(x) - I_i(y)\| \leq l\|x - y\|; \tag{7.11}$$

(P3) $\det(\mathcal{I} + B_i) \neq 0, i \in \mathbb{Z}$;
The transition matrix $X(t,s)$ of the associated homogeneous system

$$\begin{aligned} x' &= A(t)x, \\ \Delta x|_{t \neq \theta_i} &= B_i x, \end{aligned} \tag{7.12}$$

(P4) satisfies the following inequality

$$\|X(t,s)\| \leq K\mathrm{e}^{-\gamma(t-s)}, s \leq t, \tag{7.13}$$

where K and γ are positive real numbers;

(P5) $Kl[\frac{1}{\gamma} + \frac{\mathrm{e}^{\kappa\gamma}}{1-\mathrm{e}^{-\kappa\gamma}}] < 1$;

(P6) there exists a positive number H such that

$$\sup_{-\infty<t<+\infty, \|x\|<H} \|f(t,x)\| + \sup_{-\infty<i<+\infty, \|x\|<H} \|I_i(x)\| = M < +\infty,$$

and

$$KM[\frac{1}{\gamma} + \frac{\mathrm{e}^{\kappa\gamma}}{1-\mathrm{e}^{-\kappa\gamma}}] < H;$$

(P7) $\sup_{\mathbb{R}} \|A(t)\| + \sup_{\mathbb{Z}} \|B_i\| = N < +\infty$;

(P8) $-\gamma + Kl + \frac{\ln(1+Kl)}{\kappa} < 0$.

Since (7.10) belongs to the class of equations (5.38), the discussion made in Sect. 5.8 on extension of solutions of (5.38) is also valid for (7.10). Consequently, each solution $x(t) = x(t, t_0, x_0), (t_0, x_0) \in \mathbb{R} \times \mathbb{R}^n$ of (7.10) exists and is unique on $\mathbb{R}$.

Theorem 7.1.4. *If conditions $(P1) - (P7)$ are valid, then:*

(1) there exists a unique bounded solution $\phi(t) \in \mathcal{PC}^1(\mathbb{R}, \theta)$ of (7.10), and $\|\phi(t)\| < H$ for all $t \in \mathbb{R}$;

(2) if in addition condition $(P8)$ is fulfilled then $\phi(t)$ is a globally asymptotically stable solution;

(3) if (7.10) is an (ω, p)-periodic system, then $\phi(t)$ is an ω-periodic solution.

Proof. The proof falls naturally into following three parts.

1. Let us introduce a set of functions $\mathcal{H} = \{x \in \mathcal{PC}^1(\mathbb{R}, \theta) : \|x(t)\| < H, t \in \mathbb{R}\}$. Define in $\mathcal{H}$ an operator E such that, if $\xi \in \mathcal{H}$, then

$$E\xi(t) = \int_{-\infty}^{t} X(t,\tau) f(\tau,\xi) d\tau + \sum_{\theta_i < t} X(t,\theta_i+) I_i(\xi(\theta_i)). \qquad (7.14)$$

Let us show that $E : \mathcal{H} \to \mathcal{H}$. Indeed, it is easy to see that $\xi \in \mathcal{PC}^1(\mathbb{R}, \theta)$. See, for example, the proof of Theorem 2.4.1. Moreover, we have that

$$\|E\xi(t)\| \le \int_{-\infty}^{t} K e^{-\gamma(t-\tau)} M d\tau + \sum_{\theta_i < t} K e^{-\gamma(t-\theta_i)} M < H.$$

Show now, that E is a contraction. If $\xi_1(t), \xi_2(t) \in \mathcal{H}$, then

$$\|E\xi_1(t) - E\xi_2(t)\| \le \int_{-\infty}^{t} K l e^{-\gamma(t-\tau)} \|\xi_1(\tau) - \xi_2(\tau)\| d\tau +$$

$$\sum_{\theta_i < t} K l e^{-\gamma(t-\theta_i)} \|\xi_1(\theta_i) - \xi_2(\theta_i)\| <$$

$$K l [\frac{1}{\gamma} + \frac{e^{\gamma}}{1 - e^{-\kappa\gamma}}] sup_{\mathbb{R}} \|\xi_1(\tau) - \xi_2(\tau)\|.$$

That is E is a contraction and the sequence $\xi^k(t), \xi^0(t) \equiv 0, \xi^{k+1}(t) = E\xi^k(t), k = 0, 1, \ldots,$ converges to a unique bounded solution $\phi(t)$ of system (7.10) such that $\|\phi(t)\| < H$.

2. Assume that condition $(P8)$ besides $(P1) - -(P7)$ is valid, and $\psi(t)$ is another solution of system (7.10). Then

$$\phi(t) = X(t,t_0)\phi(t_0) + \int_{t_0}^{t} X(t,\tau) f(\tau,\phi(\tau)) d\tau + \sum_{t_0 \le \theta_i < t} X(t,\theta_i+) I_i(\phi(\theta_i))$$

and

$$\psi(t) = X(t,t_0)\psi(t_0) + \int_{t_0}^{t} X(t,\tau) f(\tau,\psi(\tau)) d\tau + \sum_{t_0 \le \theta_i < t} X(t,\theta_i+) I_i(\psi(\theta_i)).$$

Now, we have that

$$\|\phi(t) - \psi(t)\| \le K e^{-\gamma(t-t_0)} \|\phi(t_0) - \psi(t_0)\| + \int_{t_0}^{t} K l e^{-\gamma(t-\tau)} \|\phi(\tau) - \psi(\tau)\| d\tau +$$

$$\sum_{t_0 \le \theta_i < t} K l e^{-\gamma(t-\theta_i)} \|\phi(\theta_i) - \psi(\theta_i)\|.$$

Denote by $u(t) = \|\phi(t) - \psi(t)\|e^{\gamma t}$ in the last formula and apply Lemma 2.5.1 to obtain

$$\|\phi(t) - \psi(t)\| \leq Ke^{\kappa\gamma}e^{-(\gamma - Kl - \frac{\ln(1+Kl)}{\kappa})(t-t_0)}\|\phi(t_0) - \psi(t_0)\|.$$

The stability is proved. □

3. Assume that conditions $(P1) - -(P7)$ are valid, (7.10) is an (ω, p)-periodic system and prove, very similarly to Theorem 5.9.2, periodicity of the bounded solution $\phi(t)$, yourself.
   The theorem is proved. □

Let us consider other conditions:

(S1) each interval $[id, (i+1)d), i \in \mathbb{Z}$, where d is a positive fixed number, contains at most one element of the sequence θ;

(S2) $Kl[\frac{1}{\gamma} + \frac{e^{\gamma d}}{1-e^{-\gamma d}}] < 1$;

(S3) there exists a positive number H such that

$$\sup_{-\infty<t<+\infty, \|x\|<H} \|f(t,x)\| + \sup_{-\infty<i<+\infty, \|x\|<H} \|I_i(x)\| = M < +\infty,$$

and

$$KM[\frac{1}{\gamma} + \frac{e^{\gamma d}}{1-e^{-\gamma d}}] < H;$$

(S4) $-\gamma + Kl + \frac{\ln(1+Kl)}{d} < 0.$

We may formulate a new theorem, useful in Chap. 10, which can be proved similarly to the last assertion.

Theorem 7.1.5. *Assume that conditions* $(P2) - (P4), (P7), (S1)$–$(S3)$, *are valid. Then:*

(1) there exists a unique bounded solution $\phi(t)$ *of (7.10), and* $\|\phi(t)\| < H$ *for all* $t \in \mathbb{R}$;

(2) if additionally condition $(S4)$ *is fulfilled then* $\phi(t)$ *is an asymptotically stable solution;*

(3) if (7.10) is an (ω, p)*-periodic system, then* $\phi(t)$ *is an* ω*-periodic solution.*

Now, introduce the following assumptions:

(Q10) $K[\frac{M}{\gamma} + \frac{c(l)e^{2\gamma\kappa}}{1-e^{-\frac{\gamma\kappa}{2}}}] + b(l) < H$;

(Q11) $Kl[\frac{1}{\gamma} + \frac{k(l)e^{2\gamma\kappa}}{1-e^{-\frac{\gamma\kappa}{2}}}] < 1$;

(Q12) $-\gamma + Kl + \frac{\ln(1+Klk(l))}{\kappa} < 0,$

and consider the following assertion.

Theorem 7.1.6. *If conditions* $(P4), (Q1)-(Q5), (Q10), (Q11)$, *are valid, then:*

(1) there exists a unique bounded solution $\phi(t)$ *of (5.38), and* $\|\phi(t)\| < H$ *for all* $t \in \mathbb{R}$;

(2) if in addition condition $(Q12)$ *is fulfilled then* $\phi(t)$ *is an asymptotically stable solution;*

(3) if (5.38) is an (ω, p)*-periodic system, then* $\phi(t)$ *is an* ω*-periodic solution.*

The proof of the last theorem follows immediately Theorem 7.1.4 and the results of Sect. 5.8. Moreover, assertions (1), (2), (3) of the last theorem can be considered as corollaries of the results in Sects. 5.8, 5.9.

7.2 The Critical Case

Consider the (ω, p)-periodic system (7.1), assuming this time that we have a critical case. That is, the associated linear homogeneous system (7.4) admits a maximal set of linearly independent ω-periodic solutions $\phi_j(t), j = 1, 2, \ldots, k, 0 < k \le n$. Then, by Theorem 4.3.2, the system adjoint to (7.4),

$$\begin{aligned} y' &= -A^T(t)y, \\ \Delta y|_{t=\theta_i} &= -(\mathcal{I} + B_i^T)^{-1} B_i^T y, \end{aligned} \tag{7.15}$$

(where T denotes transposition), has k linearly independent ω-periodic solutions $\psi_j(t), j = 1, 2, \ldots, k$. Compose these solutions as the $n \times k$ matrix $H_1(t)$.

Suppose that the following condition is satisfied,

$$\int_0^\omega H_1^T(t) f(t) dt + \sum_{i=1}^p H_1^T(\theta_i) I_i = 0. \tag{7.16}$$

Then, by Theorem 4.3.3, (7.2) has the family of ω-periodic solutions $x(t, \alpha) = \alpha_1\phi_1(t) + \alpha_2\phi_2(t) + \ldots + \alpha_k\phi_k(t) + \phi_0(t)$, where $\phi_0(t)$ is an ω-periodic particular solution of (7.2).

We assume that the smoothness of system (7.1) is of higher order. That is, the matrix $A(t)$ and the function $f(t)$ are $l-1$ times, $l \ge 2$, continuously differentiable in ϵ-neighborhoods of the points $\theta_i, i \in \mathbb{Z}$. Denote the union of the neighborhoods as G_ϵ and assume that $\phi \in C^{(l-1,l,l)}(G_\epsilon \times \mathbb{R}^n \times (-\mu_0, \mu_0)) \cap C^{(0,l,l)}(\mathbb{R} \times \mathbb{R}^n \times (-\mu_0, \mu_0)), \tau_i, \Psi_i \in C^{(l,l)}(\mathbb{R}^n \times (-\mu_0, \mu_0))$.

Theorem 7.2.1. *Suppose system (7.1) satisfies the conditions discussed, (7.2) admits a family of* ω*-periodic solutions* $x(t, \alpha)$. *Let* α_0 *be a solution of the equation* $h(\alpha) = 0$, *where*

$$\begin{aligned} h(\alpha) = \int_0^\omega H_1^T(t)\phi(t, x(t, \alpha), 0)dt + \sum_{i=1}^p H_1^T(\theta_i)\{\Psi_i(x(\theta_i, \alpha), 0) + \\ \tau_i(x(\theta_i, \alpha), 0)[((\mathcal{I} + B_i)A(\theta_i) - A(\theta_i)B_i)x(\theta_i, \alpha) - A(\theta_i)I_i]\}, \end{aligned}$$

such that

$$\det[\frac{\partial h}{\partial \alpha}|_{\alpha=\alpha_0}] \neq 0.$$

Then system (7.1) has an ω-periodic solution, if $|\mu|$ is sufficiently small. The solution converges in the B-topology to $x(t, \alpha_0)$ as $\mu \to 0$.

Proof. To prove the assertion, we will use the B-equivalence method. Consider the system of ordinary differential equations

$$x' = A(t)x + f(t) + \mu\phi(t, x, \mu), \tag{7.17}$$

which is the part of system (7.1).

Fix $i \in \mathbb{Z}$ and $x \in \mathbb{R}^n$. Let $x_0(t)$ be the solution of (7.17) such that $x_0(\theta_i) = x$, and ξ_i be a solution of the equation $t = \theta_i + \mu\tau_i(x_0(t))$. That is, ξ_i is the meeting moment of $x_0(t)$ with Γ_i. Let $x_1(t)$ be the solution of system (7.17) with the initial condition $x_1(\xi_i) = B_i x_0(\xi_i) + I_i + \mu\Psi_i(, \mu)x_0(\xi_i)$. Consider the following map:

$$\begin{aligned}
J_i(x, \mu) = {} & (\mathcal{I} + B_i)\int_{\theta_i}^{\xi_i}(A(\tau)x_0(\tau) + f(\tau) + \mu\phi(\tau, x_0(\tau), \mu))d\tau + \\
& + I_i + \mu\Psi_i(x(\xi_i), \mu) + \int_{\xi_i}^{\theta_i}(A(\tau)x_0(\tau) + f(\tau) + \mu\phi(\tau, x_0(\tau), \mu))d\tau) + \\
& + I_i + \int_{\xi_i}^{\theta_i}(A(\tau)x_1(\tau) + f(\tau) + \mu\phi(\tau, x_1(\tau), \mu))d\tau.
\end{aligned} \tag{7.18}$$

We can verify that $J_i(x, \mu) = \mu\Omega_i(x, \mu)$, where $\Omega_i(x, \mu)$ is a continuously differentiable function such that $\Omega_i(x, 0) = \Psi_i(x, 0) + (\tau_i(x, 0)[((\mathcal{I} + B_i)A(\theta_i) - A(\theta_i)B_i)x - A(\theta_i)I_i]$. One can show that systems (7.1) and

$$\begin{aligned}
y' &= A(t)y + f(t) + \mu\phi(t, y, \mu), \\
\Delta y|_{t=\theta_i} &= B_i y + I_i + \mu\Omega_i(y, \mu),
\end{aligned} \tag{7.19}$$

are B-equivalent in $\mathbb{R} \times \mathbb{R}^n \times (-\mu_0, \mu_0)$. That is, the problem of existence of periodic solutions of system (7.1) can be reduced to the problem for (7.19) with fixed moments of impulsive action.

Complete the matrix $H_1(t)$ by solutions $\psi_j, j = k + 1, \ldots, n$, of the adjoint system to build a fundamental matrix of solutions $H(t)$. Substitute $z = H^T(0)y$ in (7.19) and obtain the system

$$\begin{aligned}
z' &= P(t)z + g(t) + \mu\mathcal{F}(t, z, \mu), \\
\Delta z|_{t=\theta_i} &= S_i z + K_i + \mu\mathcal{O}_i(z, \mu),
\end{aligned} \tag{7.20}$$

where

$$P(t) = H^T(0)A(t)H^T(0)^{-1}, g(t) = H^T(0)f(t),$$

$$S_i = H^T(0)B_iH^T(0)^{-1}, K_i = H^T(0)I_i,$$

$$\mathcal{F}(t,y,\mu) = H^T(0)\phi(t, H^T(0)^{-1}y,\mu), \mathcal{O}_i(y,\mu) = H^T(0)\Omega_i(H^T(0)^{-1}y,\mu).$$

Denote $z(t,\alpha) = H^T(0)x(t,\alpha), \beta = (\beta_{k+1},\dots,\beta_n)$ and let $v(t) = z(t,\alpha,\beta)$ be a solution of system (7.20) with initial condition $v(o) = z(0,\alpha) + (0,\beta)$. Further, let $L(t) = H^{-1}(0)H(t), L_1(t) = H^{-1}(0)H_1(t)$, $L_2(t)$ be the matrix composed of the entries of the last $n-k$ columns and $n-k$ rows of $L(t)$, and $L_3(t)$ be the matrix composed of the entries of the last $n-k$ rows of $L(t)$. Let

$$U(\alpha,\beta,\mu) = \int_0^\omega L_1^T(t)\mathcal{F}(t,v(t),\mu)dt + \sum_{i=1}^p L_1^T(\theta_i)\mathcal{O}_i(v(\theta_i),\mu),$$

$$V(\alpha,\beta,\mu) = (L_2^T(\omega) - \mathcal{I})\beta - \mu\int_0^\omega L_3^T(t)\mathcal{F}(t,v(t),\mu)dt - \mu\sum_{i=1}^p L_3^T(\theta_i)\mathcal{O}_i(v(\theta_i),\mu).$$

Theorem 4.3.3 implies that the ω-periodicity condition for the solution $v(t)$ has the form of the following two equations:

$$U(\alpha,\beta,\mu) = 0, \tag{7.21}$$

$$V(\alpha,\beta,\mu) = 0. \tag{7.22}$$

If in (7.22) one takes $\mu = 0$, we obtain $\beta = 0$, and then (7.21) has the form

$$U(\alpha,0,0) = \int_0^\omega L_1^T(t)\mathcal{F}(t,z(t,\alpha),0)dt + \sum_{i=1}^p L_1^T(\theta_i)\mathcal{O}(z(\theta_i,\alpha),0) = 0. \tag{7.23}$$

Let $\alpha_0 = (\alpha_1^0,\dots,\alpha_k^0)$ be a solution of (7.23). Since the function U has continuous partial derivatives with respect to $\alpha_j, j = 1,\dots,k$, in a sufficiently small neighborhood of the point $(\alpha_0, 0,)$, it follows under the assumption:

$$\det[\frac{\partial U}{\partial \alpha}|_{\alpha=\alpha_0}] \neq 0$$

that the system of equations (7.21), (7.22) is solvable with respect to α and β so that the functions $\alpha_j(\mu)$ and $\beta_s(\mu), j = 1,\dots,k, s = k+1,\dots,n$, are continuous and $\alpha_j(\mu) \to \alpha_j^0, \beta_s(\mu) \to 0$ as $\mu \to 0$.

Thus, we have established that for sufficiently small $|\mu|$ system, (7.20) admits an ω-periodic solution, which converges to the solution $x(t,\alpha_0)$ of system (7.2), as $\mu \to 0$. The solution has the form $x(t,\alpha_0) = \alpha_1^0\phi_1(t) + \dots + \alpha_r^0\phi_r(t)$, where $\phi_1(t),\dots,\phi_r(t)$, are linearly independent ω-periodic solutions of (7.2).

To complete the proof, one should apply the equivalence of (7.20) and (7.1). The theorem is proved. □

Next, we discuss the problem of asymptotic representation of the periodic solutions, whose existence has been proved in the last theorem.

Theorem 7.2.2. *Assume that system (7.1) satisfies all conditions of the last theorem, and it is smooth with order $l \geq 3$. Then, if $|\mu|$ is sufficiently small, the periodic solution $x(t)$ of this system admits the following B-asymptotic representation:*

$$x(t) = x(t, \alpha_0) + \mu x_1(t) + \ldots + \mu^{l-1} x_l(t, \mu), \tag{7.24}$$

which is valid for all $t \in \mathbb{R}$, except those $t \in \widehat{(\theta_i, \eta_i]}, i \in \mathbb{Z}$, where η_i are points of discontinuity of $x(t)$, such that $\eta_i \to \theta_i$ as $\mu \to 0$, for all $i \in \mathbb{Z}$. The functions $x_j, j = 1, \ldots, l$, belong to $\mathcal{PC}_\omega(\mathbb{R}, \theta)$.

Proof. Present the ω-periodic solution $y(t)$ of (7.19) in the form $y(t) = x(t, \alpha_0) + \mu\zeta(t, \mu)$. One can easily check that the function ζ satisfies

$$\begin{aligned} \zeta' &= A(t)\zeta + \phi(t, x(t, \alpha_0) + \mu\zeta, \mu), \\ \Delta\zeta|_{t=\theta_i} &= B_i\zeta + \Omega_i(x(\theta_i, \alpha_0) + \mu\zeta, \mu). \end{aligned} \tag{7.25}$$

We shall show that the last system has an ω-periodic solution if $|\mu|$ is sufficiently small. Denote $e(t) = \phi(t, x(t, \alpha_0), 0), m_i = \Omega_i(x(\theta_i, \alpha_0), 0), \pi(t, \zeta, \mu) = \mu^{-1}[\phi(t, x(t, \alpha_0) + \mu\zeta, \mu) - \phi(t, x(t, \alpha_0), 0)], \Pi_i(\zeta, \mu) = \mu^{-1}[\Omega_i(x(\theta_i, \alpha_0) + \mu\zeta, \mu) - \Omega_i(x(\theta_i, \alpha_0), 0)]$. Then ζ is a solution of the following system:

$$\begin{aligned} \zeta' &= A(t)\zeta + e(t) + \mu\pi(t, \zeta, \mu), \\ \Delta\zeta|_{t=\theta_i} &= B_i\zeta + m_i + \mu\Pi_i(\zeta, \mu). \end{aligned} \tag{7.26}$$

Since $h(\alpha_0) \neq 0$, the system

$$\begin{aligned} \zeta' &= A(t)\zeta + e(t), \\ \Delta\zeta|_{t=\theta_i} &= B_i\zeta + m_i, \end{aligned} \tag{7.27}$$

admits r-parametric family of ω-periodic solutions $x_1(t, \bar{\alpha}) = \bar{\alpha}_1\phi_1 + \ldots + \bar{\alpha}_r\phi_r + \phi_0$, where $\phi_0(t)$ is a particular solution of (7.27). Hence, by Theorem 7.2.1, the problem is reduced to the investigation of the expression

$$v_1(\bar{\alpha}) = \int_0^\omega H_1^T(t)\pi(t, x_1(t, \bar{\alpha}), 0)dt + \sum_{i=1}^p H_1^T(\theta_i)\Pi_i(x_1(\theta_i, \bar{\alpha}), 0). \tag{7.28}$$

It is a linear equation with respect to $\bar{\alpha} = (\bar{\alpha}_1, \ldots, \bar{\alpha}_r)$. The matrix of coefficients of the system is $\frac{\partial h}{\partial \alpha}|_{\alpha=\alpha_0}$. Indeed, we have

$$\pi(t,\zeta,0) = \phi_\mu(t,x(t,\alpha_0)+\mu\zeta,\mu)|_{\mu=0} = \sum_{j=1}^{n} \phi_{x_j}(t,x(t,\alpha_0),0)\zeta_j + \phi_\mu(t,x(t,\alpha_0),0),$$

$$\Pi_i(\zeta,0) = \sum_{j=1}^{n} \Omega_{ix_j}(x(\theta_i,\alpha_0),0)\zeta_j + \Omega_{i\mu}(x(\theta_i,\alpha_0),0).$$

Consequently, (7.28) can be written as

$$\int_0^\omega H_1^T(t) \sum_{j=1}^{n} \phi_{x_j}(t,x(t,\alpha_0),0)(\sum_{j=1}^{n} \alpha_j\phi_j$$
$$+\phi_0)dt + \int_0^\omega H_1^T(t)\phi_\mu(t,x(t,\alpha)_0,0)dt +$$

$$\sum_{i=1}^{p} H_1^T(\theta_i) \sum_{j=1}^{n} \Omega_{ix_j}(x(\theta_i,\alpha_0),0)(\sum_{j=1}^{n} \alpha_j\phi_j(\theta_i) + \phi_0(\theta_i))$$
$$+\sum_{i=1}^{p} H_1^T(\theta_i)\Omega_{i\mu}(x(\theta_i,\alpha_0),0).$$

The last expression proves the assertion. Then the equation $v_1(\bar{\alpha}) = 0$ has a unique solution $\bar{\alpha}_0$, and $\det(v_{1\alpha}(\bar{\alpha}_0)) \neq 0$ is true. Thus for system (7.27), all conditions of the last theorem are valid, and consequently, $\zeta(t,\mu) = x_1(t,\bar{\alpha}_0) + \zeta_1(t,\mu)$, where ζ_1 is convergent to 0 as $\mu \to 0$. It follows that denoting $x_0(t) = x(t,\alpha_0), x_1(t) = x_1(t,\bar{\alpha}_0)$, one can obtain $y(t) = x_0(t) + \mu x_1(t) + \mu\zeta_1(t,\mu)$. One can show that ζ_1 is a solution of the system

$$\begin{aligned} \zeta' &= A(t)\zeta + e(t) + \mu\pi(t,x_1(t,\bar{\alpha}_0) + \mu\zeta,\mu), \\ \Delta\zeta|_{t=\theta_i} &= B_i\zeta + m_i + \mu\Pi_i(x_1(\theta_i,\bar{\alpha}_0) + \mu\zeta,\mu). \end{aligned} \quad (7.29)$$

Next, assuming that $l \geq 4$, we can check that $\zeta_1 = x_2(t,\tilde{\alpha}_0) + \zeta_2(t,\mu)$, where $\zeta_2 \to 0$, as $\mu \to 0$, and x_2 is a solution of the (ω, p)-periodic equation

$$\begin{aligned} x' &= A(t)x + \pi_1(t,x_1(t),0), \\ \Delta x|_{t=\theta_i} &= B_i x + m_i + \Pi_i^1(x_1(\theta_i),0), \end{aligned} \quad (7.30)$$

$\tilde{\alpha}_0$ is a solution of the equation $v_2(\alpha) = 0$, where

$$v_2(\alpha) = \int_0^\omega H_1^T(t)\pi_1(t,x_2(t,\alpha),0)dt + \sum_{i=1}^{p} H_1^T(\theta_i)\Psi_i^1(x_1(\theta_i,\alpha),0),$$

$$\pi_1 = \mu^{-1}[\pi(t, x_1 + \mu x, \mu) - \pi(t, x_1(t), 0)], |pi_i^1$$
$$= \mu^{-1}[\Pi_i(x_1(\theta_i) + \mu x, \mu) - \Pi_i(x_1(\theta_i), 0)],$$

$$x_2(t, \alpha) = \alpha_1\phi_1 + \ldots \alpha_r\phi_r + \bar{\phi},$$

and $\bar{\phi}$ is a particular periodic solution of (7.30). In this way one can obtain the representation of $y(t)$ up to l-th order. The expansion is valid for the solution $x(t)$ of B-equivalent system (7.1). The theorem is proved. □

Notes

The Poincaré method of small parameter [132], with Lyapunov's stability [104] development was applied in [105] for intensive investigation of the existence of periodic solutions. In [112], the problem of existence of periodic solutions for strongly nonlinear systems with analytic members and nonfixed moments of impulses, was considered by using method of generalized functions. Asymptotic expansion along the powers of a small parameter for solutions of differential equations with fixed moments of impulses is discussed in [47]. We have obtained the results for a quasilinear system, considering critical and noncritical cases, by using B-equivalence method. More results on the subject can be found in [23, 26, 29, 34, 36]. The approach considered in our results allows to achieve higher order approximations. Significant development of the mechanics with impacts [41, 43, 50, 56, 57, 79, 103, 109, 113, 115, 116, 121, 125–127, 130, 162] provides an opportunity for applications. In Example 7.1.1, we consider a simple mechanical model just with an illustrative goal. Some interesting results of impacts theory [113, 126, 127] form a basis for further development of the equations considered in this Chapter. The method of small parameter can be applied to synchronization of systems with discontinuities [44, 53, 81, 94, 108, 114, 129].

Chapter 8
Discontinuous Dynamical Systems

8.1 Generalities

The book [39] edited by D.V. Anosov and V.I. Arnold considers two fundamentally different dynamical systems: flows and cascades. Roughly speaking, flows are dynamical systems with continuous time and cascades are dynamical systems with discrete time. One of the most important theoretical problems is to consider *Discontinuous Dynamical Systems* (DDS). That is, the systems whose trajectories are piecewise continuous curves. Analyzing the behavior of the trajectories, we can conclude that DDS combine features of vector fields and maps. They cannot be reduced to flows or cascades but are close to flows since time is continuous. That is why we propose to call them also as *Discontinuous Flows* (DF). One must emphasize that DF are not differential equations with discontinuous right side, which often have been accepted as DDS [68]. One should also agree that nonautonomous impulsive differential equations, which were thoroughly described in previous chapters are not discontinuous flows.

Let us remind the definition of a continuous dynamical system. Denote by X a complete metric space, with a countable base, and with ρ a metric function. A dynamical system on X is defined to be a mapping $\phi : \mathbb{R} \times X \to X$, such that

1. $\phi(0, x) = x$ for all $x \in X$, (Identical property);
2. $\phi(t + s, x) = \phi(t, \phi(s, x))$ for all $x \in X$, and $t, s \in \mathbb{R}$, (Group property);
3. $\phi(t, x)$ is a continuous function.

Definitely, one may expect that systems with similar properties can be defined for processes with discontinuities. Present chapter is devoted to the problem of identification of such kind of systems, one of the most interesting and difficult problems for impulsive differential equations.

To motivate the reader, we may propose the following simple example, where an autonomous system with even linear elements is not a dynamical system.

Example 8.1.1. Let us study the motion of the following system

$$\ddot{x} + \omega^2 x = 0,$$
$$\Delta \dot{x}|_{x=x_0} = k,$$

where ω, k, and x_0 are positive constants.

M. Akhmet, *Principles of Discontinuous Dynamical Systems*,
DOI 10.1007/978-1-4419-6581-3_8, © Springer Science+Business Media, LLC 2010

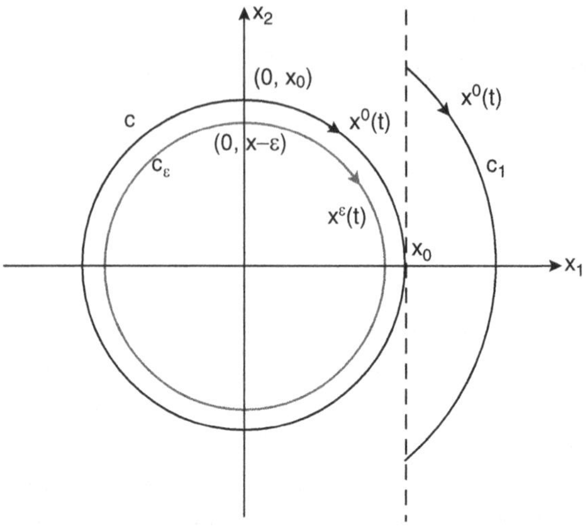

Fig. 8.1 Trajectories of the system (8.1), (8.2)

Denote $x_1 = x$ and $x_2 = \frac{1}{\omega}\dot{x}$. By using this substitution, the system can be rewritten in the form:

$$\dot{x_1} = \omega x_2, \tag{8.1}$$

$$\dot{x_2} = -\omega x_1, \quad x_1 \neq x_0,$$

$$x_2^+ = k_1 + x_2^-, \quad x_1 = x_0, \tag{8.2}$$

where $k_1 = \frac{k}{\omega}$. A solution of the system (8.1) is $x_1(t) = r \sin \omega t$ and $x_2(t) = r \cos \omega t$, where r is a fixed real number.

Let us observe the behavior of solutions of the system (8.1), (8.2) in Fig. 8.1. Consider the solution $x^0(t) = x(t, 0, (0, x_0))$. The point moves along the circle **c** until it meets the line $x_1 = x_0$ at the point $(x_0, 0)$. Then it jumps and continues to move along the arc of the circle c_1. Then, it meets the line $x_1 = x_0$ again and jumps.

One may examine that the solution is not continuous in the initial value. Indeed, let us take another solution $x^\epsilon(t) = x(t, 0, (0, x_0 - \epsilon))$ of this system, which starts at the point $(0, x_0 - \epsilon)$, where ϵ is a fixed positive real number. The solution $x^0(t)$ jumps at the point $(x_0, 0)$ and continues along the arc c_1, as explained above. However, the solution $x^\epsilon(t)$ continues its motion along the circle c_ϵ without any jump. So, as it is seen in Fig. 8.1, the distance between these two trajectories cannot be less than $\sqrt{x_0^2 - k_1^2} - x_0$, despite the initial points of these two solutions can be chosen arbitrarily close. This example demonstrates that the solution $x^0(t)$ of the system (8.1), (8.2) does not depend continuously on the initial value. Obviously, we cannot accept the system as a dynamical system. We may remark that this type of "irregularity" in models with impacts causes many interesting phenomena, for instance, collision bifurcation [51, 118].

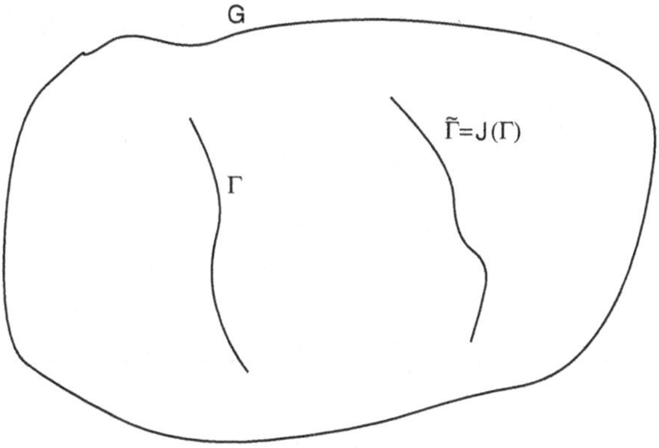

Fig. 8.2 A domain of a discontinuous dynamical system

Let $G = \bigcup G_j$ where $G_j, j = 1, 2, \ldots, m$, are disjoint open connected subsets of $\mathbb{R}^n$. Denote by G_r, an r-neighborhood of G in $\mathbb{R}^n$ for a fixed $r > 0$. Let $\Phi : G_r \longrightarrow \mathbb{R}$ be a function from $C^1(G_r)$ and assume that the surface $\Gamma = \Phi^{-1}(0)$ is a closed subset of $\bar{G}$, where $\bar{G}$ is the closure of G. Denote by Γ_r, the r-neighborhood of Γ in $\mathbb{R}^n$, and define a function $J : \Gamma_r \rightarrow G_r$, such that $J(\Gamma) \subset \bar{G}$ is a closed set. We shall need the following assumptions:

(C1) $\nabla\Phi(x) \neq 0$ for all $x \in \Gamma$;
(C2) $J \in C^1(\Gamma_r)$ and $\det[\frac{\partial J(x)}{\partial x}] \neq 0$, for all $x \in \Gamma_r$,

where $\nabla\Phi(x)$ denotes the gradient vector of Φ with respect to x. Let $\tilde{\Gamma} = J(\Gamma)$, (see Fig. 8.2), $\tilde{\Phi}(x) = \Phi(J^{-1}(x))$. One can verify that $\tilde{\Gamma} = \{x \in G |\, \tilde{\Phi}(x) = 0\}$. Condition (C1) implies that for every $x_0 \in \Gamma$ there exists a number j and a function $\varphi_{x_0}(x_1, \ldots, x_{j-1}, x_{j+1}, \ldots, x_n)$ such that Γ is the graph of the function $x_j = \varphi_{x_0}(x_1, \ldots, x_{j-1}, x_{j+1}, \ldots, x_n)$ in a neighborhood of x_0. The same is true for every $x_0 \in \tilde{\Gamma}$.

Sets Γ and $\tilde{\Gamma}$ consist of disjoint manifolds. These manifolds are with or without boundaries. We shall denote unions of all these boundaries as $\partial\Gamma$ and $\partial\tilde{\Gamma}$. One may recommend to the reader books [58,73] to recall definitions of manifolds. It is easily seen that restrictions $J|_\Gamma, \tilde{J}|_{\tilde{\Gamma}}$ are one-to-one functions.

Remark 8.1.1. It is natural to consider domains of continuous dynamical systems as connected sets [157]. Otherwise, each region of a partition can be discussed as a domain of a continuous dynamical system. A trajectory of a discontinuous dynamical system may jump from one component to another, such that only the union of the disjoint regions is a domain.

Lemma 8.1.1. $\nabla\tilde{\Phi}(x) \neq 0, \forall x \in \tilde{\Gamma}$.

Proof. We can write that $\nabla\tilde{\Phi}(x) = \nabla\Phi\left(J^{-1}(x)\right)$, and then the equality

$$\nabla\Phi\left(J^{-1}(x)\right) = \frac{\partial\Phi(y)}{\partial y}\Big|_{y=J^{-1}(x)}\frac{\partial J^{-1}(x)}{\partial x}$$

implies that

$$\nabla\Phi\left(J^{-1}(x)\right) \neq 0.$$

The lemma is proved. □

We make the following assumptions which will be needed throughout the chapter.

(C3) $f \in C^1(G_r)$,
(C4) $\Gamma \cap \tilde{\Gamma} = \emptyset$,
(C5) $\langle\nabla\Phi(x), f(x)\rangle \neq 0$ if $x \in \Gamma$,
(C6) $\langle\nabla\tilde{\Phi}(x), f(x)\rangle \neq 0$ if $x \in \tilde{\Gamma}$.

Consider the following impulsive differential equation:

$$\begin{aligned} x' &= f(x), \\ \Delta x|_{x\in\Gamma} &= W(x), \end{aligned} \tag{8.3}$$

where $W(x) = J(x) - x$, in the domain $D = \left[G \cup \Gamma \cup \tilde{\Gamma}\right] \setminus \left[\partial\Gamma \cup \partial\tilde{\Gamma}\right]$.

If $\phi(t) : I \to \mathbb{R}^n$, where I is an interval, is a solution of (8.3), then it is required that it belongs to $\mathcal{PC}(I, \theta)$, where $\theta \subset I$ is a B-sequence. The solution must satisfy $\phi'(t) = f(\phi(t))$, if $t \notin \theta$, and $\phi(\theta_i+) = J(\phi(\theta_i)), \phi(\theta_i) \in \Gamma, \phi(\theta_i+) \in \tilde{\Gamma}$, for each $\theta_i \in \theta$. Sets Γ and $\tilde{\Gamma}$ may have common points with the boundary of the domain D, and the boundary points of these sets, Γ and $\tilde{\Gamma}$, do not belong to D, as they may cause a violence of the continuous dependence on initial value. If the boundary points are in the domain, then one needs specific additional conditions. For instance, if $x \in \partial\Gamma$, then we may request $J(x) = 0$.

Now, we continue with examples, where conditions (C1)–(C6) are satisfied.

Example 8.1.2. Let us consider the following system:

$$\begin{aligned} x_1' &= -x_1 - 3x_2, \\ x_2' &= 3x_1 - x_2, \\ \Delta x_1|_{x\in\Gamma} &= x_1 \\ \Delta x_2|_{x\in\Gamma} &= x_2, \end{aligned} \tag{8.4}$$

where $x = (x_1, x_2)$, and

$$\Gamma = \left\{(x_1, x_2) | \quad x_1^2 + x_2^2 = 1, \quad x_1, x_2 \in \mathbb{R}\right\},$$

$$\tilde{\Gamma} = \left\{(x_1, x_2) | \quad x_1^2 + x_2^2 = 4, \quad x_1, x_2 \in \mathbb{R}\right\}.$$

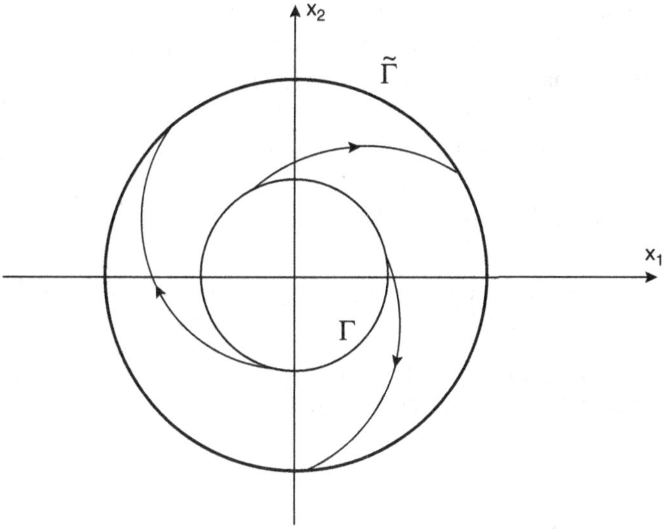

Fig. 8.3 The phase portrait of system (8.4)

Assume that $G = \{(x_1, x_2)|\ \ 1 < x_1^2 + x_2^2 < 4, \quad x_1, x_2 \in \mathbb{R}\}$. A trajectory of the system is seen in Fig. 8.3. One can easily find that $\Phi(x) = x_1^2 + x_2^2 - 1, \tilde{\Phi}(x) = x_1^2 + x_2^2 - 4, f(x) = (-x_1 - 3x_2, 3x_1 - x_2), J(x) = (2x_1, 2x_2)$. Let us check conditions (C1)–(C6). We have that $\nabla\Phi(x) = (2x_1, 2x_2) \neq 0$. So, condition (C1) is satisfied. Moreover, J, f are continuously differentiable functions and $\det[\frac{\partial J(x)}{\partial x}] = det \begin{pmatrix} 2 & 0 \\ 0 & 2 \end{pmatrix} = 4 \neq 0$, for all x. It is also obvious that $\Gamma \bigcap \tilde{\Gamma} = \emptyset$. Finally, $\langle \nabla\Phi(x), f(x)\rangle = \langle (2x_1, 2x_2), (-x_1 - 3x_2, 3x_1 - x_2)\rangle = 2\left(-x_1^2 - x_2^2\right) = -2 \neq 0$, for all $x \in \Gamma$, and $\left\langle \nabla\tilde{\Phi}(x), f(x)\right\rangle = \langle (2x_1, 2x_2), (-x_1 - 3x_2, 3x_1 - x_2)\rangle = 2\left(-x_1^2 - x_2^2\right) = -8 \neq 0$, for all $x \in \tilde{\Gamma}$. Thus, all conditions (C1)–(C6) are fulfilled.

Example 8.1.3. Let us consider the following system:

$$\begin{aligned} x_1' &= -\frac{1}{3}x_1 - 3x_2, \\ x_2' &= 3x_1 - \frac{1}{3}x_2, \\ \Delta x_1|_{x\in\Gamma} &= (2\cos\frac{\pi}{6} - 1)x_1 - 2\sin\frac{\pi}{6}x_2, \\ \Delta x_2|_{x\in\Gamma} &= 2\sin\frac{\pi}{6}x_1 + (2\cos\frac{\pi}{6} - 1)x_2. \end{aligned} \tag{8.5}$$

where $G = \mathbb{R}^2$, and $\Gamma = \{(x_1, x_2)|\ \ x_1 = x_2, \quad 0 < x_1\}$. Let us start to check conditions (C1)–(C6). One can easily find that $\tilde{\Gamma} = \{(x_1, x_2)|\ \sqrt{3}x_1 =$

$x_2, 0 < x_1\}, \Phi(x) = x_1 - x_2, \tilde{\Phi}(x) = \sqrt{3}x_1 - x_2,\ f(x) = (-\frac{1}{3}x_1 - 3x_2, 3x_1 - \frac{1}{3}x_2), J(x) = (2\cos\frac{\pi}{6}x_1 - 2\sin\frac{\pi}{6}x_2, 2\sin\frac{\pi}{6}x_1 + 2\cos\frac{\pi}{6}x_2)$. Consequently, we have that $\nabla\Phi(x) = (1,-1) \neq 0$, so, condition (C1) is satisfied. It is seen that J, f are continuously differentiable functions and $\det[\frac{\partial J(x)}{\partial x}] = \det\begin{pmatrix} 2\cos\frac{\pi}{6} & 2\sin\frac{\pi}{6} \\ 2\sin\frac{\pi}{6} & 2\cos\frac{\pi}{6} \end{pmatrix} = 4(\cos^2\frac{\pi}{6} + \sin^2\frac{\pi}{6}) = 4 \neq 0$, for all x. It is also obvious that $\Gamma \cap \tilde{\Gamma} = \emptyset$. Moreover,

$$\langle \nabla\Phi(x), f(x)\rangle = \left\langle (1,-1), (-\frac{1}{3}x_1 - 3x_2, 3x_1 - \frac{1}{3}x_2)\right\rangle = \left(\frac{-10}{3}x_1 - \frac{8}{3}x_2\right) \neq 0,$$

for all $x \in \Gamma$. The inequality $\langle\nabla\tilde{\Phi}(x), f(x)\rangle \neq 0$, for all $x \in \tilde{\Gamma}$, can be shown similarly. Thus, all conditions, (C1)–(C6) are fulfilled.

8.2 Local Existence and Uniqueness

Definition 8.2.1. A function $x(t) \in \mathcal{PC}^1(I, \theta)$, where $I \subset \mathbb{R}$ is an interval, $\theta \subset I$ is a B-sequence of discontinuity points, is said to be a solution of (8.3) if:

(i) the differential equation (8.3) is satisfied at each $t \in I \backslash \theta$ and $x'_-(\theta_i) = f(x(\theta_i)), \theta_i \in \theta$, where $x'_-(\theta_i-)$ is the left-sided derivative;
(ii) $\Delta x(\theta_i+) = W(x(\theta_i))$ for all $\theta_i \in \theta$.

Theorem 8.2.1. *Assume that conditions (C1)–(C4) hold. Then for every $x_0 \in D$ there exists an interval $(a,b) \subset \mathbb{R}, a < 0 < b$, such that a solution $x(t) = x(t, 0, x_0)$ of (8.3) exists and is unique on the interval.*

Proof. Consider the following alternative cases:

(a) Assume that $x_0 \notin \Gamma \cup \tilde{\Gamma}$. Then there exists a number $\epsilon > 0$ such that $B(x_0, \epsilon) \cap (\Gamma \cup \tilde{\Gamma}) = \emptyset$, and $B(x_0, \epsilon) \subset G$, where $B(x_0, \epsilon)$ is the ball with the center at x_0 and the radius ϵ. Therefore, by the existence and uniqueness theorem [59], $x(t)$ exists and is unique on an interval (a, b) as a solution of the system

$$y' = f(y). \tag{8.6}$$

(b) If $x_0 \in \Gamma$, then $x(0+) \in \tilde{\Gamma}$. There exists a number $\epsilon > 0$ such that $B(x(0+), \epsilon) \cap \Gamma \neq \emptyset$ and $B(x(0+), \epsilon) \subset G$. Hence, $x(t)$ can be continued at the right. Let us consider $t < 0$ now. By condition (C4), there exists a number $\epsilon > 0$ such that $B(x(0), \epsilon) \cap \tilde{\Gamma} \neq \emptyset$ and $x(t)$ can be proceeded at the left.
(c) We can discuss the case $x_0 \in \tilde{\Gamma}$ similarly to the previous one.

The uniqueness of the solution for all cases (a)–(c) follows the theorem on uniqueness of ordinary differential equations [77] and the invertability of J.

The theorem is proved. □

Since conditions (C1)–(C4) were verified in Examples 8.1.2 and 8.1.3, solutions of systems (8.4) and (8.5) locally exist and are unique.

8.3 Extension of Solutions

In this section, we will prove continuation theorems. The main results claim that every solution of (8.3) is continuable to ∞ and $-\infty$. In other words, $\mathbb{R}$ is a maximal interval of existence of each solution $x(t, 0, x_0), x_0 \in D$ of (8.3). That is, $x(t, 0, x_0) \in \mathcal{PC}(\mathbb{R})$. Illustrating examples are given, where solutions exist on $\mathbb{R}$.

Definition 8.3.1. A solution $x(t) = x(t, 0, x_0)$ of (8.3) is said to be continuable to a set $S \subset \mathbb{R}^n$ as time decreases (increases) if there exists a moment $\xi \in \mathbb{R}$, such that $\xi \leq 0$ $(\xi \geq 0)$ and $x(\xi) \in S$.

The following theorems provide sufficient conditions for the continuation of solutions of (8.3).

Theorem 8.3.1. *Assume that:*

(a) *every solution $y(t, 0, x_0), x_0 \in D$, of (8.6) is continuable to either ∞ or Γ, as time increases;*
(b) *there exists a positive number $\bar{\theta}$ such that*

$$\frac{\epsilon_x}{\sup_{B(x,\epsilon_x)} \|f(x)\|} \geq \bar{\theta},$$

for every $x \in \tilde{\Gamma}$ and all $\epsilon_x > 0$ with $B(x, \epsilon_x) \cap \Gamma = \emptyset$.

Then every solution $x(t) = x(t, 0, x_0), x_0 \in D$, of (8.3) is continuable to ∞.

Proof. Fix $x_0 \in D$ and let $x(t) = x(t, 0, x_0)$ be a solution of (8.3). Consider the following two cases.

(A) If x(t) is a continuous solution of (8.3), then it is a solution of (8.6) and is continuable to ∞.
(B) Let $x(\theta_i+) \in \tilde{\Gamma}$ for a fixed i. We set $M_x = \sup_{B(x,\epsilon_x)} \|f(x)\|$. Assume that there exists a number $\xi > \theta_i$, such that $||x(\xi) - x(\theta_i+)|| = \epsilon_{x(\theta_i+)}$ (otherwise $x(t)$ is continuable to ∞). Then

$$x(\xi) = x(\theta_i+) + \int_{\theta_i}^{\xi} f(x(s))ds,$$

and $\epsilon_{x(\theta_i+)} \leq M_{x(\theta_i+)} (\xi - \theta_i) \leq M_{x(\theta_i+)} (\theta_{i+1} - \theta_i)$, where $M_{x(\theta_i+)} > 0$ (Why?). The last inequality implies that $\theta_{i+1} - \theta_i \geq \bar{\theta}$ for all i. That is, θ_i is a sequence of β-type if $\theta_i \geq 0$. The proof is complete. □

In a similar manner, one can prove that the following theorem is valid.

Theorem 8.3.2. *Assume that:*

(a) every solution $y(t,0,x_0)$, $x_0 \in D$ *of (8.6) is continuable to either* $-\infty$ *or* $\tilde{\Gamma}$, *as time decreases;*

(b) there exists a positive number $\bar{\theta}$ *such that*

$$\frac{\epsilon_x}{\sup_{B(x,\epsilon_x)} \|f(x)\|} \geq \bar{\theta},$$

for every $x \in \Gamma$ *and all* $\epsilon_x > 0$ *with* $B(x,\epsilon_x) \cap \tilde{\Gamma} = \emptyset$.

Then, every solution $x(t) = x(t,0,x_0)$, $x_0 \in D$, *of (8.3) is continuable to* $-\infty$.

Theorems 8.3.1 and 8.3.2 imply that the following assertion is valid.

Theorem 8.3.3. *Assume that*

(a) every solution $y(t,0,x_0)$, $x_0 \in D$, *of (8.6) satisfies the following conditions:*

(a1) it is continuable to either ∞ *or* Γ, *as time increases,*
(a2) it is continuable to either $-\infty$ *or* $\tilde{\Gamma}$, *as time decreases;*

(b) there exists a positive number $\bar{\theta}$ *such that*

$$\frac{\epsilon_x}{\sup_{B(x,\epsilon_x)} \|f(x)\|} \geq \bar{\theta},$$

for every $x \in \tilde{\Gamma}$ *and all* $\epsilon_x > 0$ *with* $B(x,\epsilon_x) \cap \Gamma = \emptyset$.

(c) there exists a positive number $\tilde{\theta}$ *such that*

$$\frac{\tilde{\epsilon}_x}{\sup_{B(x,\tilde{\epsilon}_x)} \|f(x)\|} \geq \tilde{\theta},$$

for every $x \in \Gamma$ *and all* $\tilde{\epsilon}_x > 0$ *with* $B(x,\tilde{\epsilon}_x) \cap \tilde{\Gamma} = \emptyset$.

Then, every solution $x(t) = x(t,0,x_0)$, $x_0 \in D$, *of (8.3) is continuable on* $\mathbb{R}$.

Other sufficient conditions for the continuation of solutions of (8.3) are provided by the following theorems.

Theorem 8.3.4. *Assume that*

(a) every solution $y(t,0,x_0)$, $x_0 \in D$, *of (8.6) satisfies the following conditions:*

(a1) it is continuable either to ∞ *or* Γ, *as* t *increases;*
(a2) it is continuable either to $-\infty$ *or* $\tilde{\Gamma}$, *as* t *decreases;*

(b) $\sup_D \|f(x)\| < +\infty$.
(c) $\text{dist}(\Gamma, \tilde{\Gamma}) > 0$.

Then every solution $x(t,0,x_0)$, $x_0 \in D$, *of (8.3) is continuable on* $\mathbb{R}$.

Proof. Fix $x_0 \in D$ and let $x(t) = x(t, 0, x_0)$ be a solution of (8.3). According to Definition 2.1.1, we shall consider the following three cases:

(A) If x(t) is a continuous solution of (8.3), then it is a solution of (8.6) and, thus is continuable on $\mathbb{R}$.

(B) Denote by $\theta_{\max}$ and $\theta_{\min}$ the maximal and minimal elements of the set $\{\theta_i\}$, respectively. Consider $t \geq \theta_{\max}$. By the condition on J we have that $x(\theta_{\max}+) = J(x(\theta_{\max}-)) \in D$ and the solution $x(t) = y\,(t, \theta_{\max}, x(\theta_{\max}+))$, where y is a solution of (8.6) and is continuable to ∞. For $t \leq \theta_{\min}$, one can apply the same arguments to show that x(t) is continuable to $-\infty$.

(C) Consider the following three alternatives.

(c_1) If the sequence $\{\theta_i\}$ has a maximal element $\theta_{\max} \in \mathbb{R}$, but does not have a minimal one, then by using (B), it is easy to prove that $x(t)$ is continuable to ∞. Let t be decreasing. We have that

$$x(\theta_i+) = x(\theta_{i+1}) + \int_{\theta_{i+1}}^{\theta_i} f(x(s))ds. \tag{8.7}$$

Denote $\sup_D \|f(x)\| = M$ and $\operatorname{dist}(\Gamma, \tilde{\Gamma}) = \alpha$. Then (8.7) implies that $\frac{\alpha}{M} \leq (\theta_{i+1} - \theta_i)$. Hence, $\frac{\alpha}{M}(i - i_0) \geq \left(\theta_i - \theta_{i_0}\right)$, where i_0 is fixed. The last inequality shows that $\theta_i \to -\infty$ as $i \to -\infty$. Thus, $x(t)$ is continuable to $-\infty$.

(c_2) Assume that the sequence $\{\theta_i\}$ has a minimal element $\theta_{\min}$, and does not have a maximal one. Then by the arguments of (B) $x(t)$ is continuable to $-\infty$. For increasing t we have that

$$x(\theta_{i+1}) = x(\theta_i+) + \int_{\theta_i}^{\theta_{i+1}} f(x(s))ds, \tag{8.8}$$

$\frac{\alpha}{M} \leq (\theta_{i+1} - \theta_i)$ or $\frac{\alpha}{M}(i - i_0) \leq \left(\theta_i - \theta_{i_0}\right)$, where i_0 is fixed. Hence, $\theta_i \to \infty$ as $i \to \infty$. That is, $x(t)$ is continuable to ∞.

(c_3) Assume that $\{\theta_i\}$ has neither a minimal nor a maximal element. The result for this case follows (c_1) and (c_2). The proof is complete. □

Theorem 8.3.5. *Assume that*

(a) every solution $y(t, 0, x_0), x_0 \in D$, of (8.6) is continuable to either ∞ or Γ, as time increases;

(b) there exists a neighborhood S of Γ in D such that

  (b1) $\operatorname{dist}(\Gamma, \partial S) > 0$;

  (b2) $\sup_S \|f(x)\| < \infty$;

  (b3) $\tilde{\Gamma} \cap S = \emptyset$.

Then every solution $x(t) = x(t, 0, x_0), x_0 \in D$, of (8.3) is continuable to ∞.

Proof. Denote $d = \text{dist}(\Gamma, \partial \mathrm{S})$ and $M = \sup_S \|f(x)\|$. For a fixed i one can see that

$$x(\theta_{i+1}) = x(\theta_i+) + \int_{\theta_i}^{\theta_{i+1}} f(x(s))ds.$$

Condition (b3) implies that $d < \|x(\theta_{i+1}) - x(\theta_i+)\| \le M(\theta_{i+1} - \theta_i)$. Thus $\theta_{i+1} - \theta_i \ge \frac{d}{M} > 0$ for all i. Further discussion is fully analogous to that of the last theorem. □

Exercise 8.3.1. Prove the following theorem.

Theorem 8.3.6. *Assume that:*

(a) every solution $y(t, 0, x_0), x_0 \in D$, of (8.6) is continuable to either $-\infty$ or $\tilde{\Gamma}$, as time decreases,
(b) there exists a neighborhood $\tilde{S}$ of $\tilde{\Gamma}$ in D such that:

  (b1) $\text{dist}(\tilde{\Gamma}, \partial \tilde{\mathrm{S}}) > 0$;
  (b2) $\sup_{\tilde{S}} \|f(x)\| < \infty$;
  (b3) $\Gamma \cap \tilde{S} = \emptyset$.

Then, every solution $x(t) = x(t, 0, x_0), x_0 \in D$, of (8.3) is continuable to $-\infty$.

Using the conditions of both Theorems 8.3.5 and 8.3.6, one can formulate the following assertion.

Theorem 8.3.7. *Assume that:*

(a) every solution $y(t, 0, x_0), x_0 \in D$, of (8.6) satisfies the following conditions:

  (a1) it is continuable to either ∞ or Γ, as time increases;
  (a2) it is continuable to either $-\infty$ or $\tilde{\Gamma}$, as time decreases;

(b) there exist neighborhoods S and $\tilde{S}$ of Γ and $\tilde{\Gamma}$ in D, respectively, such that:

  (b1) $\text{dist}(\Gamma, \partial \mathrm{S}) > 0, \text{dist}(\tilde{\Gamma}, \partial \tilde{\mathrm{S}}) > 0$;
  (b2) $\sup_{S \cup \tilde{S}} \|f(x)\| < \infty$;
  (b3) $\tilde{\Gamma} \cap S = \emptyset, \Gamma \cap \tilde{S} = \emptyset$.

Then, every solution $x(t) = x(t, 0, x_0), x_0 \in D$, of (8.3) is continuable on $\mathbb{R}$.

Example 8.3.1. Let us consider system (8.5) and study the extension property. The differential equation in the system is a linear one, consequently, each solution of this equation is continuable to ∞, since maximal interval of existence is $\mathbb{R}$. The first condition of Theorem 8.3.1 is satisfied. Let us fix an initial value $x_0 = (x_1^0, x_2^0) \in \tilde{\Gamma}$, that is $\sqrt{3}x_1^0 = x_2^0$. Then, one can easily evaluate the distance between Γ and x_0

$$\text{dist}(\mathrm{x}_0, \Gamma) = \frac{|\mathrm{x}_1^0 - \mathrm{x}_2^0|}{\sqrt{2}} = \frac{\sqrt{3} - 1}{\sqrt{2}} |\mathrm{x}_1^0| = \frac{\sqrt{3} - 1}{2\sqrt{2}} \|\mathrm{x}_0\|.$$

Fix

$$\epsilon_{x_0} = \frac{\sqrt{3}-1}{2\sqrt{2}} \|x_0\| \tag{8.9}$$

and take any $x \in B\left(x_0, \epsilon_{x_0}\right)$, then

$$\|x\| < \epsilon_{x_0} + \|x_0\|. \tag{8.10}$$

Substituting (8.9) into (8.10), one can conclude that

$$\|x\| < \left[\frac{\sqrt{3}-1+2\sqrt{2}}{2\sqrt{2}}\right] \|x_0\|.$$

Computing the norm of the function f in this ball, we get that

$$\|f(x)\| \leq \frac{\sqrt{41}}{6}\left[\sqrt{3}-1+2\sqrt{2}\right] \|x_0\| = M_{x_0}.$$

By easy calculation,

$$\inf \frac{\epsilon_x}{M_x} = \frac{\frac{\sqrt{3}-1}{2\sqrt{2}} \|x_0\|}{\frac{\sqrt{41}}{6}\left[\sqrt{3}-1+2\sqrt{2}\right] \|x_0\|} = \frac{3\left(\sqrt{3}-1\right)}{\sqrt{82}\left(\sqrt{3}-1+2\sqrt{2}\right)} > 0.$$

We can see, now, that condition (*b*) is valid. Thus, all conditions of Theorem 8.3.1 are satisfied, and every solution of system (8.5) is continuable to ∞. The continuation of solutions for decreasing t can be shown by using Theorem 8.3.2.

Example 8.3.2. Let us examine system (8.4). The domain of this system is $D = \{(x_1, x_2)|\ \ 1 \leq x_1^2 + x_2^2 \leq 4, \quad x_1, x_2 \in \mathbb{R}\}$. Manifolds Γ and $\tilde{\Gamma}$ are boundaries of this ring. They are circles with radii 1 and 2, and $\operatorname{dist}(\tilde{\Gamma}, \Gamma) = 1$, respectively.

The differential equation in (8.4) is a linear system with constant coefficients, and one can determine that all solutions are continuable to Γ, as time increases, and are continuable to $\tilde{\Gamma}$, as time decreases. Hence, the first condition of Theorem 8.3.4 is satisfied.

Moreover,

$$\|f(x)\| = \sqrt{(-x_1-3x_2)^2 + (3x_1-x_2)^2} = \sqrt{10}\sqrt{x_1^2+x_2^2}, \tag{8.11}$$

and

$$\sup_D \|f(x)\| = 2\sqrt{10} < \infty.$$

Since, all conditions of Theorem 8.3.4 are satisfied, every solution of system (8.4) is continuable on $\mathbb{R}$.

Example 8.3.3. Consider the following impulsive autonomous system:

$$\begin{aligned}
x_1' &= -2x_1 - 3x_2,\\
x_2' &= 3x_1 - 2x_2,\\
\Delta x_1|_{x\in\Gamma} &= (2\cos\frac{\pi}{6} - 1)x_1 - 2\sin\frac{\pi}{6}x_2,\\
\Delta x_2|_{x\in\Gamma} &= 2\sin\frac{\pi}{6}x_1 + (2\cos\frac{\pi}{6} - 1)x_2,
\end{aligned} \tag{8.12}$$

where manifolds of discontinuity are

$$\Gamma = \left\{(x_1, x_2)| \quad x_1 = \sqrt{3}x_2, \quad \frac{1}{2} < x_2 < \frac{3}{2}\right\}$$

and

$$\tilde{\Gamma} = \left\{(x_1, x_2)| \quad \sqrt{3}x_1 = x_2, \quad 1 < x_1 < 3\right\}.$$

Domain $D = \mathbb{R}^2 \backslash \left\{\left(\frac{\sqrt{3}}{2}, \frac{1}{2}\right), \left(\frac{3\sqrt{3}}{2}, \frac{3}{2}\right), \left(1, \sqrt{3}\right), \left(3, 3\sqrt{3}\right)\right\}$. Let us look for sufficient conditions of Theorem 8.3.5 to indicate continuation of solutions of the system (8.12) for increasing t. The differential equation in (8.12) is a linear system and maximal interval of existence is $\mathbb{R}$, so each solution of the differential equation is continuable to ∞ as time increases. Hence, the first condition is satisfied. While dealing with other conditions, we prefer to use both polar and Cartesian coordinates. First, let us define an auxiliary set S in polar coordinates (see Fig. 8.4),

$$S = \left\{(\rho, \theta) \mid \quad \frac{9}{10} < \rho < \frac{21}{10}, \quad \frac{\pi}{12} < \theta < \frac{\pi}{4}\right\}.$$

One can easily see that $\Gamma \subset S$ and $\tilde{\Gamma} \cap S = \emptyset$. The distance between Γ and ∂S, is the minimum of the following two numbers: the distance between Γ and the arc $\gamma = \{(\rho, \theta)|\rho = \frac{9}{10}, \frac{\pi}{12} < \theta < \frac{\pi}{4}\}$; the distance between Γ and the line $\ell = \{(\rho, \theta)|\frac{9}{10} < \rho < \frac{21}{10}, \theta = \frac{\pi}{4}\}$. One can find that $dist(\Gamma, \gamma) = \frac{1}{10}$. Next, let us write the equation of the line in Cartesian coordinates as

$$\ell = \left\{(x_1, x_2) \mid \quad x_1 = x_2, \ x_1, x_2 \in \mathbb{R}^+\right\}.$$

To find $dist(\Gamma, \ell)$, it is sufficient to find out the distance between the line ℓ and the points $A\left(\frac{\sqrt{3}}{2}, \frac{1}{2}\right)$ and $B\left(\frac{3\sqrt{3}}{2}, \frac{3}{2}\right)$,

$$dist\,(\ell, A) = \frac{\left|\frac{\sqrt{3}}{2} - \frac{1}{2}\right|}{\sqrt{1+1}} = \frac{\sqrt{3}-1}{2\sqrt{2}}, \qquad dist\,(\ell, B) = \frac{\left|\frac{3\sqrt{3}}{2} - \frac{3}{2}\right|}{\sqrt{1+1}} = \frac{3\sqrt{3}-3}{2\sqrt{2}}.$$

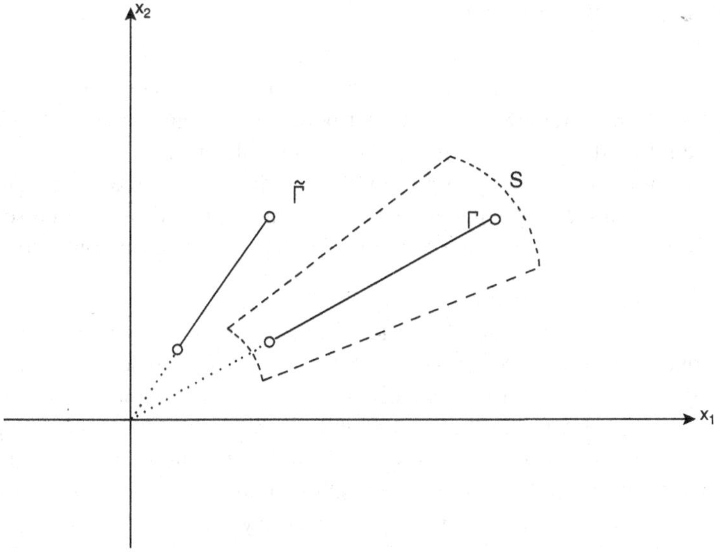

Fig. 8.4 Manifolds Γ, $\tilde{\Gamma}$, and an auxiliary set S

Then, distance between Γ and the surface ∂S is

$$dist(\Gamma, \partial S) = \frac{1}{10}.$$

Now, we take the norm of the function $f(x)$,

$$\|f(x)\| = \sqrt{4+9}\sqrt{x_1^2 + x_2^2}.$$

Since

$$\frac{9}{10} < \sqrt{x_1^2 + x_2^2} < \frac{21}{10},$$

$$\sup_S \|f(x)\| = \frac{21\sqrt{13}}{10}.$$

Thus, all conditions of Theorem 8.3.5 are satisfied, and every solution of system (8.12) is continuable to ∞.

Exercise 8.3.2. Prove, by using Theorem 8.3.6, that all solutions of system (8.12) are continuable to $-\infty$.

8.4 The Group Property

In the previous sections of the chapter, we have dealt with existence and uniqueness of solutions of the system (8.3), and furthermore, we have given the conditions that are sufficient for all solutions of (8.3) to be continuable on $\mathbb{R}$.

Now, we may discuss the group property, which is one of the most significant properties of dynamical systems and one of the most difficult for the present discussion. Next example shows that even in a simple case the group property can be violated.

Example 8.4.1. Let us consider the system (8.4), where we only replace the set G by a new one $G = \{(x_1, x_2)| \quad x_1^2 + x_2^2 > 1, \quad x_1, x_2 \in \mathbb{R}\}$. To demonstrate that the group property is not valid for all solutions, we use Fig. 8.5. Consider a trajectory, which starts at x_0 and reaches the point P at some positive moment t. Moving back it could not return to x_0, for decreasing t, because of the discontinuity set $\tilde{\Gamma}$. That is, equality $x(-t, 0, x(t, 0, x_0)) = x_0$, which is a consequence of the property is not true for all moments of time. Hence, the property is not valid for the system. It is obvious, also, that uniqueness of solutions is not true in this case, and it is not surprising, as it is known that the group property and the uniqueness are strongly related to each other.

The last example shows that specific conditions to guarantee the group property should be found.

The following condition is one of the most needed in this chapter.

(C7) (a) for every $x \in \Gamma$ there exists $\epsilon_x > 0$ such that $\mathrm{sign}\Phi(x)$ is a constant function in $[B(x, \epsilon_x) \cap G] \backslash \Gamma$;
(b) for every $x \in \tilde{\Gamma}$ there exists $\epsilon_x > 0$ such that $\mathrm{sign}\tilde{\Phi}(x)$ is a constant function in $[B(x, \epsilon_x) \cap G] \backslash \tilde{\Gamma}$.

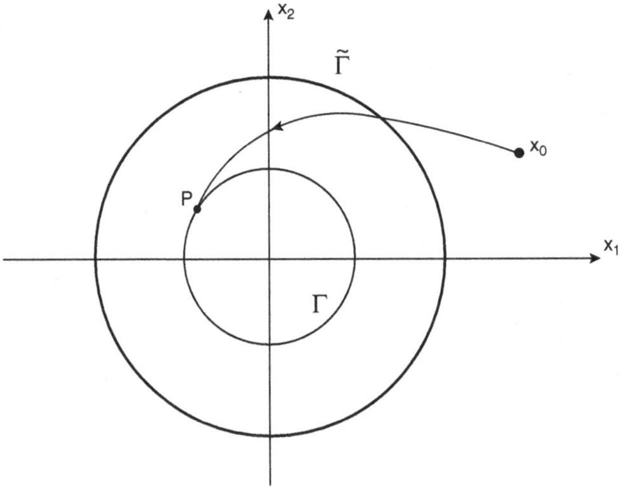

Fig. 8.5 The trajectory of Example 8.4.1

Lemma 8.4.1. *Assume that (C1)–(C7) hold and* $y(t) : (-\alpha, \alpha) \to \mathbb{R}^n$, $\alpha > 0$, *is a solution of (8.6). Then* $y(0) \notin \Gamma$ *and* $y(0) \notin \tilde{\Gamma}$.

Proof. Assume, on the contrary, that $y(0) = y_0 \in \Gamma$. We have that

$$\Phi(y(t)) = \Phi(y(t)) - \Phi(y_0) = \langle \nabla\Phi(y_0), y(t) - y_0\rangle + o(||y(t) - y_0||) =$$

$$\langle \nabla\Phi(y_0), f(y_0)t + o(|t|)\rangle + o(||f(y_0)||t + o(|t|)) = \langle \nabla\Phi(y(0)), f(y(0))\rangle\, t + o(|t|).$$

By condition (C7) function $\operatorname{sign}\Phi(y(t))$ has a constant value for sufficiently small $|t|$, and by condition (C4) the value of $\langle \nabla\Phi(y(0)), f(y(0))\rangle$ is not zero. This contradiction proves our lemma for Γ. The proof for $\tilde{\Gamma}$ is similar. □

Lemma 8.4.2. *Assume that (C1)–(C7) hold. Then* $x(-t, 0, x(t, 0, x_0)) = x_0$ *for all* $x_0 \in D, t \in \mathbb{R}$.

Proof. Consider $t > 0$. If the set $\{\theta_i\}$ is empty, then the proof follows immediately the assertion for continuous dynamical systems [39]. One can see that it remains to check the validity of $x(-\theta_i, 0, x(\theta_i+)) = x(\theta_i)$ for all i, and the condition $x(-\theta_1, 0, x(\theta_1, 0, x_0)) = x_0$. The first one is obvious because of invertability of J. Let us consider the second one. Denote $x(t) = x(t, 0, x_0)$, $\tilde{x}(t) = x(t, 0, x(\theta_1))$. Since $x(\theta_1) \in \Gamma$, then by (C4), the solution $\tilde{x}$ moves along the trajectory of (8.6) for decreasing t, and it cannot meet $\tilde{\Gamma}$ if $t > -\theta_1$. Indeed, assume on the contrary that there exists moment θ, $-\theta_1 < \theta < 0$, where $\tilde{x}$ intersects $\tilde{\Gamma}$. Then $\tilde{x}(\theta+) = x(\theta + \theta_1)$. We have obtained a contradiction to Lemma 8.4.1 since $x(t + \theta + \theta_1)$ is the solution of (8.6) in a neighborhood of $t = 0$. If $t < 0$, the proof is very similar to that of $t > 0$, and the proof with $t = 0$ is primitive. The lemma is proved. □

Let us continue with the following auxiliary result.

Consider a solution $x(t) : \mathbb{R} \to \mathbb{R}^n$ of (8.3). Let $\{\theta_i\}$ be the sequence of discontinuity points of $x(t)$. Fix $\bar{\theta} \in \mathbb{R}$ and introduce a function $\psi(t) = x(t + \bar{\theta})$.

Lemma 8.4.3. *The sequence* $\{\theta_i - \bar{\theta}\}$ *is a set of all solutions of the equation*

$$\Phi(\psi(t)) = 0. \tag{8.13}$$

Proof. We have $\Phi(\psi((\theta_i - \bar{\theta})) = \Phi(x((\theta_i - \bar{\theta}) + \bar{\theta})) = \Phi(x(\theta_i)) = 0$. Assume that $t = \varphi$ is a solution of (8.13), then $\Phi(x(\varphi + \bar{\theta})) = \Phi(\psi(\varphi)) = 0$. That is, $\varphi + \bar{\theta}$ is one of the numbers $\{\theta_i\}$. Let $\varphi + \bar{\theta} = \theta_j$, then $\varphi = \theta_j - \bar{\theta}$. The lemma is proved. □

Lemma 8.4.4. *If* $x(t) : \mathbb{R} \to \mathbb{R}^n$ *is a solution of (8.3), then* $x(t + \bar{\theta})$, $\bar{\theta} \in \mathbb{R}$, *is also a solution of (8.3).*

Proof. From Lemma 8.4.3, it follows that $\psi = x(t + \bar{\theta})$ is a continuous function on the interval $(\theta_i - \bar{\theta}, \theta_{i+1} - \bar{\theta}]$, $i \in \mathbb{Z}$. Fix $i \in \mathbb{Z}$, and consider $t \in (\theta_i - \bar{\theta}, \theta_{i+1} - \bar{\theta}]$.

We have that $t + \bar{\theta} \in (\theta_i, \theta_{i+1}]$ and one can verify that $\psi'(t) = f(\psi(t))$. That is, (8.3) is satisfied by $x(t + \theta)$.

For fixed i, we have that $\psi((\theta_i - \bar{\theta})+) = x(\theta_i+) = J(x(\theta_i)) = J(\psi(\theta_i - \bar{\theta}))$. Thus, one can see that the jump equations in (8.3) are also satisfied by $x(t + \bar{\theta})$, and this completes the proof. □

Lemmas 8.4.2 and 8.4.4 imply that the following theorem is valid. The proof of this theorem is similar to that of continuous dynamical systems [157].

Theorem 8.4.1. *Assume that conditions (C1)–(C7) are fulfilled. Then*

$$x(t_2, x(t_1, x_0)) = x(t_2 + t_1, x_0), \tag{8.14}$$

for all $t_1, t_2 \in \mathbb{R}$.

Remark 8.4.1. Since $x(0, x_0) = x_0$, one can conclude on the basis of Theorem 8.4.1 that $x(t, x_0), t \in \mathbb{R}, x_0 \in D$, defines a one-parameter group of transformations of D into itself.

Exercise 8.4.1. Verify that condition (C7) is fulfilled in Example 8.1.2, and it is not correct in Example 8.4.1.

8.5 Continuity Properties

A dependence of solutions on initial values is a very effective method to investigate various problems of dynamical systems, and we deal with the continuous dependence in this section. It is assumed that all considered solutions are continuable on $\mathbb{R}$. The next example demonstrates that the continuity property should be discussed very carefully when one is busy with nonfixed moments of discontinuity.

Example 8.5.1. Consider the autonomous system

$$\begin{aligned} x_1' &= 0, \\ x_2' &= 0, \\ \Delta x_1'|_{x \in \Gamma} &= 0, \\ \Delta x_2'|_{x \in \Gamma} &= -1, \end{aligned} \tag{8.15}$$

where $\Gamma = \{x \in \mathbb{R}^2 : x_1 = x_2\}$. Take solutions $x_0(t) = x(t, 0, (3, 3))$, and $x(t) = x(t, 0, (x_0^1, x_0^2)), x_0^1 > 3, x_0^2 < 3$, and consider them for increasing t. One can easily see that the more points (x_0^1, x_0^2) and $(3, 3)$ are close, the more the distance $||x_0(t) - x(t)||, t > 0$, is close to 1.

Fix a point $x_0 \in \Gamma \backslash \partial \Gamma$, and denote by $B(x_0, r)$ an open ball with the center at x_0 and the radius $r > 0$. By condition (C5), if r is sufficiently small, the ball is divided by the surface Γ into two connected and open regions. Denote by $b^+(x_0, r)$ the region, which $x(t, 0, x_0)$ enters as time decreases. Let $c^+(x_0, r) = (\Gamma \cap B(x_0, r)) \cup b^+(x_0, r)$. If $x_0 \notin \Gamma$, then $c^+(x_0, r) = B(x_0, r)$, where the radius r so small that $B(x_0, r) \cap \Gamma = \emptyset$. Similarly, if $x_0 \in \tilde{\Gamma} \backslash \partial \tilde{\Gamma}$ denote by $b^-(x_0, r)$ the region, which $x(t, 0, x_0)$ enters as time increases. Then write $c^-(x_0, r) = (\tilde{\Gamma} \cap B(x_0, r)) \cup b^-(x_0, r)$. We set also $c^-(x_0, r)$ equals $B(x_0, r)$, where the radius r so small that $B(x_0, r) \cap \tilde{\Gamma} = \emptyset$, if $x_0 \notin \tilde{\Gamma}$.

Let $x^0(t) : \mathbb{R} \to \mathbb{R}^n, x^0(t) = x(t, 0, x_0)$, be a solution of (8.3).

Definition 8.5.1. The solution $x^0(t)$ of (8.3) B-continuously depends on x_0 for increasing t, if to any $\epsilon > 0$ and finite interval $[0, b], 0 < b$, there corresponds $\delta > 0$ such that any other solution $x(t) = x(t, 0, \bar{x})$ of (8.3) lies in the ϵ-neighborhood of $x^0(t)$ on $[0, b]$, if $\bar{x} \in c^+(x_0, \delta)$.

Definition 8.5.2. The solution $x^0(t) : \mathbb{R} \to \mathbb{R}^n, x^0(t) = x(t, 0, x_0)$, of (8.3) B-continuously depends on x_0 for decreasing t, if to any $\epsilon > 0$ and finite interval $[a, 0], a < 0$, there corresponds $\delta > 0$ such that any other solution $x(t) = x(t, 0, \bar{x})$ of (8.3) lies in the ϵ-neighborhood of $x^0(t)$ on $[a, 0]$, if $\bar{x} \in c^-(x_0, \delta)$.

Definition 8.5.3. The solution $x^0(t) : \mathbb{R} \to \mathbb{R}^n, x^0(t) = x(t, 0, x_0)$, of (8.3) B-continuously depends on x_0 if it continuously depends on the initial value for both decreasing and increasing t.

Theorem 8.5.1. *Assume that conditions (C1)–(C6) are satisfied. Then each solution $x^0(t) = x(t, 0, x_0), x_0 \in D$, of (8.3) continuously depends on x_0.*

Proof. We consider a particular case with a finite interval $[0, b]$, and the points of discontinuity $\theta_i, i = 1, \ldots, m$, of the solution $x^0(t)$ in the interval such that $0 < \theta_1 < \cdots < \theta_m < b$. Moreover, we assume that $t = 0$, and $t = b$ are not the moments of discontinuity. All other cases can be considered similarly.

Fix a positive number α. Let $F_\alpha = \{(t, x) | t \in [0, b], \|x - x^0(t)\| < \alpha\}, G_i(\alpha)$, $i = 0, 1, 2, \ldots, m + 1$, be α-neighborhoods of points $(0, x_0)$, $(\theta_i, x(\theta_i))$, $i = 1, 2, \ldots, m$, $(b, x^0(b))$ in $\mathbb{R} \times \mathbb{R}^n$, respectively, and $\bar{G}_i(\alpha), i = 1, 2, \ldots, k$, be an α-neighborhood of the point $(\theta_i, x^0(\theta_i+))$. Write

$$G^\alpha = F_\alpha \cup \left(\cup_{i=0}^{m+1} G_i(\alpha)\right) \cup \left(\cup_{i=1}^{m} \bar{G}_i(\alpha)\right).$$

Take α sufficiently small so that $G^\alpha \subset \mathbb{R} \times D$. Fix $\epsilon, 0 < \epsilon < \alpha$.

1. In view of the continuity of solutions [77], there exists $\bar{\delta}_m, 0 < \bar{\delta}_m < \epsilon$, such that every solution $x_m(t)$ of (8.6), which starts in $\bar{G}_m(\bar{\delta}_m)$, is continuable to $t = b$, does not intersect Γ, and

$$\|x_m(t) - x^0(t)\| < \epsilon,$$

for all t from the common domain of $x_m(t)$ and $x^0(t)$.
2. By continuity of J there exists $0 < \delta_m < \epsilon$, such that $(\kappa, x) \in G_m(\delta_m)$ implies $(\kappa, x + J(x)) \in \bar{G}_m(\bar{\delta}_m) \cap D$.

3. The continuity theorem yields that there exists $\bar{\delta}_{m-1}, 0 < \bar{\delta}_{m-1} < \epsilon$, such that a solution $x_{m-1}(t)$ of (8.6), which starts in $\bar{G}_{m-1}(\bar{\delta}_{m-1})$, intersects Γ in $G_m(\delta_m)$ (we continue the solution $x_{m-1}(t)$ only to the moment of the intersection) and $\left\|x_{m-1}(t) - x^0(t)\right\| < \epsilon$ for all t from the common domain of $x_{m-1}(t)$ and $x^0(t)$.

Continuing the process for $m-2, m-3, \ldots, 1$, one can obtain a sequence of families of solutions of (8.6) $x_i(t)$, and corresponding numbers $\delta_i, \bar{\delta}_i, i = 1, 2, \ldots, m$. Finally, we find a number $\delta, 0 < \delta < \epsilon$, such that each solution $x_0(t)$, which starts in $G_0(\delta)$ intersects Γ in $G_1(\delta_1)$, if t increases, and satisfies $||x_0(t) - x^0(t)|| < \epsilon$ if t is from the common domain of $x_0(t)$ and $x^0(t)$. Thus, if one chooses a solution $x(t) = x(t, 0, \bar{x}), \bar{x} \in G_0(\delta)$, of (8.3), then it coincides over the first interval of continuity, except possibly, the δ_1-neighborhood of θ_1, with one of the solutions $x_0(t)$. Then on the interval $[\theta_1, \theta_2]$ it coincides with one of the solutions $x_1(t)$, except possibly, the δ_1-neighborhood of θ_1 and the δ_2-neighborhood of θ_2, etc. Finally, one can see that the integral curve of $x(t)$ belongs to G^ϵ, it has exactly m meeting points with Γ, $\theta_i^1, i = 1, 2, \ldots, m$, $\left|\theta_i^1 - \theta_i\right| < \epsilon$ for all i, and it is continuable to $t = b$. The theorem is proved. □

8.6 B-Equivalence

In this section, we construct an auxiliary system of differential equations with impulses at fixed moments, a B-equivalent system, for equations (8.3). One have to emphasize that B-equivalence plays less general role for autonomous impulsive systems than for nonautonomous equations. In this part of the manuscript, we specify a B-equivalent system around a solution of equations (8.3).

First, we need to introduce two maps, which will be used throughout the rest of the chapter. Fix $\kappa \in \mathbb{R}$. Denote by $x(t) = x(t, \kappa, x)$ a solution of (8.6), $\tau = \tau(x)$ the moment of the meeting of $x(t)$ with the surface Γ.

Lemma 8.6.1. $\tau(x) \in C^1$.

Proof. Differentiating $\Phi(x(\tau, \kappa, x)) = 0$, and using (C5) one can get that

$$\frac{\partial \Phi(x(\tau, \kappa, x))}{\partial \tau} = \frac{\partial \Phi(x(\tau, \kappa, x))}{\partial x} \frac{dx(t)}{dt}\bigg|_{t=\tau} = \frac{\partial \Phi(x(\tau, \kappa, x))}{\partial x} f(x(\tau, \kappa, x)) \neq 0$$

Now, the proof follows immediately the implicit function theorem. □

Corollary 8.6.1. *$\tau(x)$ is a continuous function.*

Let $x_1 = x(t, \tau, x(\tau)) + J(x(\tau))$ be another solution of (8.6). Define the map $\Psi(x) = x_1(\kappa)$.

Similarly to Lemma 8.6.1, one can show that the following assertion is valid.

Lemma 8.6.2. $\Psi(x) \in C^1$

Consider a solution $x^0(t) : [a,b] \to R^n, a \le 0 \le b$, of (8.3). Assume that all discontinuity points $\theta_i, i = -k, \dots, -1, 1, \dots, m$, are interior points of $[a,b]$. That is, $a < \theta_{-k}$ and $\theta_m < b$.

The following system of differential equations with impulses at fixed moments, which are points of discontinuity of $x^0(t)$, is very important in the sequel:

$$y' = f(y),$$
$$\Delta y|_{t=\theta_i} = W_i(y(\theta_i)). \tag{8.16}$$

The function f is the same as in (8.3) and maps $W_i, -k \le i \le m$, will be defined below. There exists a positive number r, such that r-neighborhoods $G_i(r)$ of $(\theta_i, x^0(\theta_i))$ do not intersect each other. In view of (C5), one can suppose that r is sufficiently small so that every solution of (8.6) which starts in $G_i(r)$ intersects Γ in $G_i(r)$ as t increases or decreases.

Fix $i = -k, \dots, m$ and let $\xi(t) = x(t, \theta_i, x), (\theta_i, x) \in G_i(r)$, be a solution of (8.6), $\tau_i = \tau_i(x)$ the meeting time of $\xi(t)$ with Γ and $\psi(t) = x(t, \tau_i, \xi(\tau_i) + J(\xi(\tau_i))$ another solution of (8.6). One should mention that $|\tau_i(x) - \theta_i| = O(r)$. Denote $W_i(x) = \psi(\theta_i) - x$. One can see that

$$W_i(x) = \int_{\theta_i}^{\tau_i} f(\xi(s))ds + J(x + \int_{\theta_i}^{\tau_i} f(\xi(s))ds) + \int_{\tau_i}^{\theta_i} f(\psi(s))ds \tag{8.17}$$

is a map of an intersection of the plane $t = \theta_i$ with $G_i(r)$ into the plane $t = \theta_i$. The functions $W_i, -k \le i \le m$, are obtained by using the map Ψ, which has been defined above in this section. Hence, Lemma 8.6.2 implies that all W_i are continuously differentiable maps.

Let us introduce the following sets: $F_r = \{(t,x) | t \in [a,b], \|x - x^0(t)\| < r\}$, and $\bar{G}_i(r), i = -k, \dots, m$, an r-neighborhood of the point $(\theta_i, x^0(\theta_i+))$. Write

$$G^r = F_r \cup \left(\cup_{i=-k}^m G_i(r)\right) \cup \left(\cup_{i=-k}^m \bar{G}_i(r)\right).$$

Take r sufficiently small so that $G^r \subset \mathbb{R} \times D$. Denote by $G(h)$ a h-neighborhood of $x^0(0)$.

Definition 8.6.1. Systems (8.3) and (8.16) are said to be B-equivalent in G^r if there exists $h > 0$, such that:

1. for every solution $x(t)$ of (8.3) such that $x(0) \in G(h)$, the integral curve of $x(t)$ belongs to G^r and there exists a solution $y(t) = y(t, 0, x(0))$ of (8.16) which satisfies

$$x(t) = y(t), t \in [a,b] \setminus \cup_{i=-k}^m \widehat{(\tau_i, \theta_i]}, \tag{8.18}$$

where τ_i are moments of discontinuity of $x(t)$. Particularly:

$$x(\theta_i) = \begin{cases} y(\theta_i), & \text{if } \theta_i \le \tau_i, \\ y(\theta_i^+), & \text{otherwise,} \end{cases}$$
$$y(\tau_i) = \begin{cases} x(\tau_i), & \text{if } \theta_i \ge \tau_i, \\ x(\tau_i^+), & \text{otherwise.} \end{cases} \tag{8.19}$$

2. Conversely, if (8.16) has a solution $y(t) = y(t, 0, x(0)), x(0) \in G(h)$, then there exists a solution $x(t) = x(t, 0, x(0))$ of (8.3) which has an integral curve in G^r, and (8.18) holds.

The following assertion follows immediately (8.17).

Lemma 8.6.3. *$x^0(t)$ is a solution of (8.3) and (8.16) simultaneously.*

Theorem 8.6.1. *Assume that conditions (C1)–(C6) are fulfilled. Then systems (8.3) and (8.16) are B-equivalent in G^r if r is sufficiently small.*

Proof. Assume that $r > 0$ is small so that W_i, $i = -k, \ldots, -1, 1, \ldots, m$, are defined. Let us check only the first condition of Definition 8.6.1 because that of the second one is analogous. Theorem 8.5.1 implies that there exists a small $h, 0 < h < r$, such that if $\|\bar{x} - x_0\| < h$ and $\bar{x} \in D$, then the solution $x(t) = x(t, 0, \bar{x})$ belongs to G^r. Assume that h is sufficiently small so that $x(t)$ has exactly $m + k$ moments of discontinuity $t = \tau_i, i = -k, \ldots, -1, 1, \ldots, m$. Without loss of generality, we suppose that $\theta_i > \tau_i$ for all i. It is obvious that we need only to prove the theorem for $[0, b]$, because for $[a, 0]$, the proof is similar. Consider the solution $y(t) = y(t, 0, x(0))$ of (8.16). By the theorem on existence and uniqueness [77] the equality

$$x(t) = y(t) \tag{8.20}$$

is valid on $[0, \tau_1]$. Since $(\tau_1, x(\tau_1)) \in G^r$ we see that

$$y(\theta_1+) = y(\tau_1) + \int_{\tau_1}^{\theta_1} f(y(s))ds + W_i(y(\theta_1)) \tag{8.21}$$

is defined and

$$x(\theta_1) = x(\tau_1) + J(x(\tau_1)) + \int_{\tau_1}^{\theta_1} f(x(s))ds. \tag{8.22}$$

Using (8.20)–(8.22) one can obtain that

$$y(\theta_1+) = x(\tau_1) + \int_{\tau_1}^{\theta_1} f(y(s))ds + \int_{\theta_1}^{\tau_1} f(y(s))ds +$$

$$J(y(\tau_1)) + \int_{\tau_1}^{\theta_1} f(x(s))ds = x(\theta_1).$$

Now, defining $x(t)$ and $y(t)$ as solutions of (8.6) with a common initial value $x(\theta_1)$, one can see that $x(t) = y(t), t \in (\theta_1, \tau_2]$. Continuing in the same manner for all $t \in [0, b]$ one can show that $y(t)$ is continuable to $t = b$ and (8.18) holds. Moreover, it is easily seen that for sufficiently small h, the integral curve of $y(t)$ belongs to G_r. The theorem is proved. □

8.7 Differentiability Properties

Let us consider derivatives of functions $\tau_i(x)$, $W_i(x), i = -k, \ldots, -1, 1 \ldots, m$, which were described in Sect. 8.6. We start with derivatives of $\tau_i(x)$. One should emphasize that $\tau_i, i = -k, \ldots, -1, 1 \ldots, m$ are maps, which are defined by the map τ in Sect. 8.6 with $\kappa = \theta_i, i = -k, \ldots, -1, 1 \ldots, m$. The equalities $\Phi(x(\tau_i(x))) = 0$ imply that

$$\Phi_x(x^0(\theta_i)) f(x^0(\theta_i)) d\tau_i + \sum_{k=1}^{n} \Phi_x(x^0(\theta_i)) \frac{\partial x^0(\theta_i)}{\partial x_k} dx_k = 0.$$

Using the last expression, one can obtain that

$$\frac{\partial \tau_i(x^0(\theta_i))}{\partial x_j} = -\frac{\Phi_x(x^0(\theta_i)) \frac{\partial x^0(\theta_i)}{\partial x_j}}{\Phi_x(x^0(\theta_i)) f(x^0(\theta_i))}. \tag{8.23}$$

Similarly, for W_i the following expression is valid:

$$\frac{\partial W_i(x^0(\theta_i))}{\partial x_j} = f \frac{\partial \tau_i}{\partial x_j} + \frac{\partial J}{\partial x}(e_j + f \frac{\partial \tau_i}{\partial x_j}) - f^+ \frac{\partial \tau_i}{\partial x_j}. \tag{8.24}$$

Thus, formulas (8.23) and (8.24) provide evaluations of the derivatives.

It is known that $x^0(t) : [a, b] \to \mathbb{R}^n$ is the solution of (8.3) and (8.16). Moreover, systems (8.3) and (8.16) are B-equivalent in G^r and there exists $\delta \in R, \delta > 0$, such that every solution which starts in $c^+(x_0, r)$ is continuable to $t = b$. Without loss of generality, assume that all points of discontinuity of $x^0(t)$ are interior points of $[a, b]$. Denote by $x^j(t), j = 1, 2, \ldots, n$, solutions of (8.3) such that $x^j(0) = x_0 + \xi e_j = (x_1^0, x_2^0, \ldots, x_{j-1}^0, x_j^0 + \xi, x_{j+1}^0, \ldots, x_n^0), \xi \in \mathbb{R}$, and let θ_i^j be the moments of discontinuity of $x^j(t)$. By Theorem 8.5.1, a solution $x^j(t), j = 1, 2, \ldots, n$, is defined on $[a, b]$ if $x_0 + \xi e_j$ belongs to $c^+(x_0, \delta)$ and $c^-(x_0, \delta)$ with sufficiently small δ.

Definition 8.7.1. The solution $x^0(t)$ is B-differentiable with respect to $x_j^0, j = 1, 2, \ldots, n$, on $[a, b]$ if for all $x_0 + \xi e_j$, which belong to $c^+(x_0, \delta)$ and $c^-(x_0, \delta)$ with sufficiently small δ it is true that:

A) there exist constants $\nu_{ij}, i = -k, \ldots, -1, 1, \ldots, m$, such that

$$\theta_i^j - \theta_i = \nu_{ij}\xi + o(|\xi|); \tag{8.25}$$

B) for all $t \in [a,b] \setminus \cup_{i=-k}^{m} \widehat{(\theta_i, \theta_i^j]}$, the following equality is satisfied:

$$x^j(t) - x^0(t) = u_j(t)\xi + o(|\xi|), \tag{8.26}$$

where $u_j(t) \in \mathcal{PC}([a,b], \theta])$.
The pair $\{u_j, \{\nu_{ij}\}_i\}$ is said to be a B-derivative of $x^0(t)$ with respect to x_j^0 on $[a,b]$.

Lemma 8.7.1. *Assume that conditions (C1)–(C6) hold. Then the solution $x^0(t)$ of (8.16) has B-derivatives with respect to $x_j^0, j = 1,2,\ldots,n$, on $[a,b]$. Moreover, u_j is a solution of the linear system*

$$\begin{aligned} \frac{du}{dt} &= f_x(x^0(t))u, \\ \Delta u|_{t=\theta_i} &= W_{ix}(x^0(\theta_i))u(\theta_i), \end{aligned} \tag{8.27}$$

with $u(0) = e_j$, and constants $\nu_{ij} = 0$, for all i.

Proof. We shall prove the lemma with respect to x_1^0. Let $y_1(t) = y(t, 0, x_0 + \xi e_1)$. By the theorem on differentiability with respect to parameters [77] we have that $y_1(t) - x^0(t) = u_1(t)\xi + \rho(\xi)$, $\rho(\xi) = o(|\xi|)$, for all $t \in [0, \theta_1]$. Particularly, $y_1(\theta_1) - x^0(\theta_1) = u_1(\theta_1)\xi + \rho(|\xi|)$. Then $y_1(\theta_1+) - x^0(\theta_1+) = W_1(y_1(\theta_1)) - W_1(x^0(\theta_1)) = W_{1x}(x^0(\theta_1))[u_1(\theta_1)\xi + \rho(\xi)] + \bar{\rho}_1(\xi)$. Since $\bar{\rho}_1 = o(|\xi|)$, we have that $y_1(\theta_1+) - x^0(\theta_1+) = u_1(\theta_1+)\xi + \tilde{\rho}_1(\xi)$, where $\tilde{\rho}_1 = o(|\xi|)$. Denote by $U(t), U(\theta_1) = \mathcal{I}$, the fundamental matrix of the system $u'(t) = f_x(x^0(t))$. Using the theorem from [60, 77] one can obtain that for all $t \in (\theta_1, \theta_2]$ the following relation is true $y_1(t) - x^0(t) = U(t)(y_1(\theta_1+) - x^0(\theta_1+)) + \rho(y_1(\theta_1+) - x^0(\theta_1+)) = U(t)u_1(\theta_m+)\xi + \rho_2(\xi) = u_1(t)\xi + \rho_2(\xi)$, where $\rho_2 = o(|\xi|)$. Continuing the process we can prove that (8.26) is valid. Formula (8.25) is trivial. The lemma is proved. □

Theorem 8.7.1. *Assume that conditions (C1)–(C6) are satisfied. Then the solution $x^0(t)$ of (8.3) has the B-derivative with respect to $x_j^0, j = 1,2,\ldots,n$, on $[a,b]$. Moreover, the derivative $(u_j(t), \{\nu_{ij}\})$ is a solution of the variational system*

$$\begin{aligned} \frac{du}{dt} &= f_x(x^0(t))u, \\ \Delta u|_{t=\theta_i} &= W_{ix}(x^0(\theta_i))u(\theta_i), \\ \nu_{ij} &= -\frac{\Phi_x u(\theta_i)}{\Phi_x f}, \end{aligned} \tag{8.28}$$

with $u(0) = e_j$.

The last theorem follows immediately Theorem 8.6.1, Lemma 8.7.1, and formulas (8.23), (8.24).

Remark 8.7.1. Higher order differentiability of DDS is considered in [3].

8.8 Conclusion

Let $D \subset \mathbb{R}^n$ be a set, which is described for system (8.3) in the introductory part of this chapter.

Definition 8.8.1. A B-smooth discontinuous flow is a map $\phi : \mathbb{R} \times D \to D$, which satisfies the following properties:

(I) The *group property*:

(i) $\phi(0, x) : D \to D$ is the identity;
(ii) $\phi(t, \phi(s, x)) = \phi(t + s, x)$ is valid for all $t, s \in \mathbb{R}$ and $x \in D$.

(II) $\phi(t, x) \in \mathcal{PC}^1(\mathbb{R})$ for each fixed $x \in D$.
(III) $\phi(t, x)$ is B-differentiable in $x \in D$ on $[a, b] \subset \mathbb{R}$ for each a, b such that the discontinuity points of $\phi(t, x)$ are interior points of $[a, b]$.

Remark 8.8.1. One can see that system (8.3) defines a B-smooth discontinuous flow provided that (C1)–(C7) and the conditions of one of the extension theorems are fulfilled.

Let us weaken the smoothness condition to obtain the definition of a discontinuous flow.

Definition 8.8.2. A B-flow is a map $\phi : \mathbb{R} \times D \to D$, which satisfies the property (I) of Definition 8.8.1 and the following conditions:

(IV) $\phi(t, x) \in \mathcal{PC}(\mathbb{R})$, for each fixed $x \in D$, and $\phi(\theta_i, x) \in \Gamma, \quad \phi(\theta_i+, x) \in \tilde{\Gamma}$ for every discontinuity point.
(V) $\phi(t, x)$ is B-continuous in x on each finite and closed interval.

Remark 8.8.2. Comparing definitions of the B-differentiability and the B-continuity, one can conclude that every B-smooth discontinuous flow is a B-flow.

Exercise 8.8.1. Use the discontinuous dynamics to arrange a partition of D.

8.9 Examples

Example 8.9.1. Consider the following impulsive differential system:

$$
\begin{aligned}
x_1' &= \alpha x_1 - \beta x_2,\\
x_2' &= \beta x_1 + \alpha x_2,
\end{aligned}
$$

$$\Delta x_1|_{x\in\Gamma} = (\sqrt{3}+1)x_1 - x_2$$
$$\Delta x_2|_{x\in\Gamma} = x_1 + (\sqrt{3}+1)x_2, \tag{8.29}$$

where $\Gamma = \{(x_1, x_2)|\, x_2 = \frac{1}{2}x_1, x_1 > 0\}$, $\tilde{\Gamma} = \{(x_1, x_2)|x_2 = \frac{\sqrt{3}}{2}x_1, x_1 > 0\}$, constants α, β are positive. One can see that $\Phi(x) = x_2 - \frac{1}{2}x_1$, $f(x) = (\alpha x_1 - \beta x_2, \beta x_1 + \alpha x_2)$, $J(x) = (\sqrt{3}x_1 - x_2, x_1 + \sqrt{3}x_2)$. We assume that

$$D = \mathbb{R}^2 \backslash \left[\left\{(x_1, x_2)|\quad \frac{1}{2}x_1 < x_2 < \frac{\sqrt{3}}{2}x_1, \quad x_1 > 0\right\} \cup (0,0)\right].$$

One can verify that the functions and the sets satisfy (C1)–(C7). Let us check if the conditions of Theorem 8.3.3 are fulfilled. Fix $x \in \tilde{\Gamma}$. Then $\text{dist}(x, \Gamma) = \frac{1}{2}||x||$ and

$$||f(x)|| = \sqrt{(\alpha x_1 - \beta x_2)^2 + (\beta x_1 + \alpha x_2)^2} = \sqrt{\alpha^2+\beta^2}||x||.$$

Thus

$$\sup_{B(x,\epsilon_x)} ||f|| = \sqrt{\alpha^2+\beta^2}(||x|| + \frac{1}{2}||x||) = \frac{3}{2}\sqrt{\alpha^2+\beta^2}||x||,$$

and

$$\inf_{\tilde{\Gamma}\times(0,\infty)} \frac{\epsilon_x}{\sup_{B(x,\epsilon_x)} ||f||} = \frac{2}{3\sqrt{\alpha^2+\beta^2}} > 0.$$

Hence, all conditions of a discontinuous flow are fulfilled.

Example 8.9.2. Consider the following model of a simple neural nets from [123]. We have modified it according to the system (8.3).

$$x_1' = x_2,$$
$$x_2' = -\beta^2 x_1,$$
$$p' = -\gamma p + x_1 + B_0,$$
$$\Delta p|_{(x,p)\in\Gamma} = -p, \tag{8.30}$$

where $\Gamma = \{(x_1, x_2, p)|\, p = r,\, x_1^2 + \frac{x_2^2}{\beta^4} < 1\}$, $\tilde{\Gamma} = \{(x_1, x_2, p)| \quad p = 0,\, x_1^2 + \frac{x_2^2}{\beta^4} < 1\}$, $\Phi(x) = p - r, f(x) = (x_2, -\beta^2 x_1, -\gamma p + x_1 + B_0)$, $J(x) = (x_1, x_2, 0)$, $\beta,\gamma,r > 0$, are constants and $B_0 > \gamma r + 1$. We assume that $D = \{(x_1, x_2, p)|0 \le p \le r, x_1^2 + \frac{x_2^2}{\beta^4} < 1\}$. The variable p is a scalar input of a neural trigger and x_1, x_2, are other variables. The value of r is the threshold. One can verify that the functions and the sets satisfy (C1)–(C7) and the conditions of Theorem 8.3.4. That is, the system defines a B-smooth discontinuous flow.

Example 8.9.3. Let us consider the following system

$$\begin{aligned}
x_1' &= \alpha x_1 - \beta x_2, \\
x_2' &= \beta x_1 + \alpha x_2, \\
\Delta x_1|_{x\in\Gamma} &= (1+k)x_1, \\
\Delta x_2|_{x\in\Gamma} &= (1+k)x_2,
\end{aligned} \tag{8.31}$$

where $\Gamma = \{(x_1, x_2)|\, x_1^2 + x_2^2 = r\}, \tilde{\Gamma} = \{(x_1, x_2)|\, x_1^2 + x_2^2 = kr\}$, α, β, k are constants such that $\alpha,\ \beta < 0, 1 < k$. Assume that $D = \mathbb{R}^2$. One can see that all conditions (C1)–(C6) are valid for the system. But (C7) is not fulfilled. It is easy to see that a solution $x(t, 0, x_0)$ of (8.31), which starts outside of $\tilde{\Gamma}$, does not satisfy the condition $x(-t, 0, x(t, 0, x_0)) = x_0$ for all t. Thus (8.31) does not determine a discontinuous flow.

Notes

Apparently, T. Pavlidis [123,124], was the first, who formulated the problem of conditions for autonomous equations with discontinuities, which guarantee properties of dynamical systems. Papers [123, 124, 135, 136] contain interesting practical and theoretical ideas concerning discontinuous flows. These authors formulated some important conditions on differential equations, but not all of them were used to prove basic properties of discontinuous flows. Some ideas on the dynamical properties can be found also in [54,87,95,111].

The chapter embodies results that provide conditions for the existence of a *discontinuous flow* and a *differentiable discontinuous flow*. Concepts of B-continuous and B-differentiable dependence of solutions on initial values are applied to describe DDS and to obtain conditions for the extension of solutions and the group property. Since DF have specific smoothness of solutions we call these systems *B-differentiable discontinuous flows.* The results are due to [1]. Since some conditions of the chapter are sufficient, but not necessary, one can develop them, but we are confident that B-continuity and B-differentiability of a motion cannot be ignored in the future investigations. It is obvious that results of the chapter can be extended for smooth of higher order and analytic discontinuous dynamics.

Chapter 9
Perturbations and Hopf Bifurcation of a Discontinuous Limit Cycle

This chapter is organized in the following manner. In the first section, we give the description of the systems under consideration and prove the theorem of existence of foci and centers of the nonperturbed system. The main subject of Sect. 9.2 is foci of the perturbed equation. The noncritical case is considered. In Sect. 9.3, the problem of distinguishing between the center and the focus is solved. Bifurcation of a periodic solution is investigated in Sect. 9.4. The last section consists of examples illustrating the bifurcation theorem.

9.1 The Nonperturbed System

Denote by $< x, y >$ the dot-product of vectors $x, y \in \mathbb{R}^2$, and $||x|| =< x, x >^{\frac{1}{2}}$ the norm of a vector $x \in \mathbb{R}^2$. Moreover, let $\mathcal{R}$ be the set of all real valued constant 2×2 matrices, and $\mathcal{I} \in \mathcal{R}$ be the identity matrix.

D$_0$-system. Consider the following differential equation with impulses:

$$\begin{aligned} \frac{dx}{dt} &= Ax, \\ \Delta x|_{x \in \Gamma_0} &= B_0 x, \end{aligned} \tag{9.1}$$

where Γ_0 is a subset of $\mathbb{R}^2$, and it will be described below, $A, B_0 \in \mathcal{R}$.

The following assumptions will be needed throughout this chapter:

(C1) $\Gamma_0 = \cup_{i=1}^{p} s_i$, where p is a fixed natural number and half-lines $s_i, i = 1, 2, \ldots, p$, are defined by equations $< a^i, x >= 0$, where $a^i = (a_1^i, a_2^i)$ are constant vectors. The origin does not belong to the lines (see Fig. 9.1).

(C2)

$$A = \begin{pmatrix} \alpha & -\beta \\ \beta & \alpha \end{pmatrix},$$

where $\alpha, \beta \in \mathbb{R}, \beta \neq 0$;

M. Akhmet, *Principles of Discontinuous Dynamical Systems*,
DOI 10.1007/978-1-4419-6581-3_9,

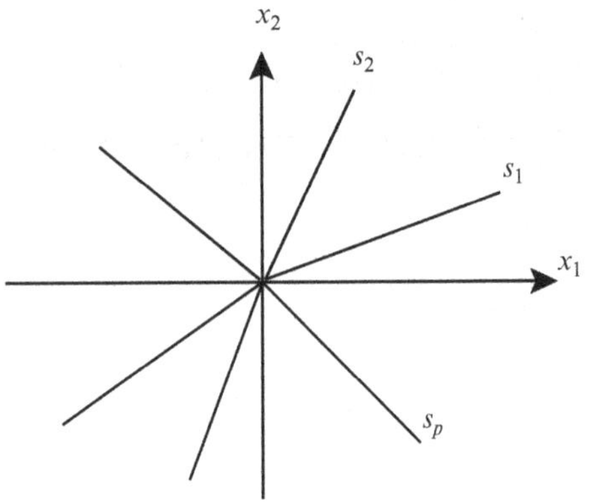

Fig. 9.1 The domain of the nonperturbed system (9.1) with a vertex which unites the straight lines $s_i, i = 1, 2, \ldots, p$

(C3) there exists a regular matrix $Q \in \mathcal{R}$ and nonnegative real numbers k and θ such that

$$B_0 = kQ \begin{pmatrix} \cos\theta & -\sin\theta \\ \sin\theta & \cos\theta \end{pmatrix} Q^{-1} - \begin{pmatrix} 1 & 0 \\ 0 & 1 \end{pmatrix};$$

We consider every angle for a point with respect to the positive half-line of the first coordinate axis. Denote $s_i^{'} = (\mathcal{I} + B_0)s_i, i = 1, 2, \ldots, p$. Let γ_i and ζ_i be angles of s_i and $s_i^{'}, i = 1, 2, \ldots, p$, respectively,

$$B_0 = \begin{pmatrix} b_{11}\, b_{12} \\ b_{21}\, b_{22} \end{pmatrix}.$$

(C4) $0 < \gamma_1 < \zeta_1 < \gamma_2 < \cdots < \gamma_p < \zeta_p < 2\pi, (b_{11} + 1)\cos\gamma_i + b_{12}\sin\gamma_i \neq 0,$ $i = 1, 2, \ldots, p.$

If conditions (C1)–(C4) hold, then (9.1) is said to be a D_0 – *system.*

Exercise 9.1.1. Verify that the origin is a unique singular point of a D_0 – system and (9.1) is not a linear system.

Exercise 9.1.2. Using the results of the last chapter, prove that D_0 – system (9.1) provides a B-smooth discontinuous flow.

If we use transformation $x_1 = r\cos(\phi), x_2 = r\sin(\phi)$ in (9.1) and exclude the time variable t, we can find that the solution $r(\phi, r_0)$ which starts at the point $(0, r_0)$, satisfies the following system:

$$\begin{aligned} &\frac{dr}{d\phi} = \lambda r, \\ &\Delta r\, |_{\phi=\gamma_i (\mathrm{mod} 2\pi)} = k_i r, \end{aligned} \tag{9.2}$$

where $\lambda = \frac{\alpha}{\beta}$, the angle-variable ϕ is ranged over the set

$$R_\phi = \cup_{i=-\infty}^{\infty}[\cup_{j=1}^{p-1}(2\pi i + \zeta_j, 2\pi i + \gamma_{j+1}] \cup (2\pi i + \zeta_p, 2\pi(i+1) + \gamma_1]]$$

and $k_i = [((b_{11}+1)\cos(\gamma_i)+b_{12}\sin(\gamma_i))^2+(b_{21}\cos(\gamma_i)+(b_{22}+1)\sin(\gamma_i))^2]^{\frac{1}{2}}-1$. Equation (9.2) is 2π-periodic, so we shall consider just the section $\phi \in [0, 2\pi]$ in what follows. That is, the system

$$\begin{aligned} \frac{dr}{d\phi} &= \lambda r, \\ \Delta r \mid_{\phi=\gamma_i} &= k_i r, \end{aligned} \tag{9.3}$$

is considered with $\phi \in [0, 2\pi]_\phi \equiv [0, 2\pi] \setminus \cup_{i=1}^{p} (\gamma_i, \zeta_i]$. System (9.3) is a sample of the time scale differential equation with transition condition [19]. We shall reduce (9.3) to an impulsive differential equation [4, 19] for the investigation's needs. Indeed, let us introduce a new variable $\psi = \phi - \sum_{0<\gamma_j<\phi} \theta_j, \theta_j = \zeta_j - \gamma_j$, with the range $[0, 2\pi - \sum_{i=1}^{p} \theta_i]$. We shall call this new variable ψ-substitution. It is easy to check that upon ψ-substitution the solution $r(\phi, r_0)$ satisfies the following impulsive equation:

$$\begin{aligned} \frac{dr}{d\psi} &= \lambda r, \\ \Delta r \mid_{\psi=\delta_j} &= k_j r, \end{aligned} \tag{9.4}$$

where $\delta_j = \gamma_j - \sum_{0<\gamma_i<\gamma_j} \theta_i$. Solving the last impulsive system and using the inverse of ψ-substitution, one can obtain that the solution $r(\phi, r_0)$ of (9.2) has the form

$$r(\phi, r_0) = \exp(\lambda(\phi - \sum_{0<\gamma_i<\phi} \theta_i)) \prod_{0<\gamma_i<\phi} (1 + k_i) r_0, \tag{9.5}$$

if $\phi \in [0, 2\pi]_\phi$.

Denote

$$q = \exp(\lambda(2\pi - \sum_{i=1}^{p} \theta_i)) \prod_{i=1}^{p} (1 + k_i). \tag{9.6}$$

Applying the Poincaré return map $r(2\pi, r_0)$ to (9.5) one can obtain that the following theorem follows.

Theorem 9.1.1. *If*

(1) $q = 1$, then the origin is a center and all solutions of (9.1) are periodic with period $T = (2\pi - \sum_{i=1}^{p} \theta_i)\beta^{-1}$;
(2) $q < 1$, then the origin is a stable focus;
(3) $q > 1$, then the origin is an unstable focus of D_0 − system.

9.2 The Perturbed System

Theorem 2.4.1 of the last section implies that if conditions (C1)–(C4) are valid, then each trajectory of (9.1) either spirals to the origin or is a discontinuous cycle. Moreover, if the trajectory spirals to the origin then it spirals to infinity, too. That is, the asymptotic behavior of the trajectory is very similar to the behavior of trajectories of the planar linear system of ordinary differential equations with constant coefficients [59,77]. In what follows, we will consider how a perturbation may change the phase portrait of the system.

D-system. Let us consider the following equation:

$$\begin{aligned} \frac{dx}{dt} &= Ax + f(x), \\ \Delta x|_{x \in \Gamma} &= B(x)x, \end{aligned} \tag{9.7}$$

in a neighborhood G of the origin.

The following is the list of conditions assumed for this system:

(C5) $\Gamma = \cup_{i=1}^{p} l_i$ is a set of curves which start at the origin and are determined by the equations $< a^i, x > + \tau_i(x) = 0, i = 1, 2, \ldots, p$. The origin does not belong to the curves (see Fig. 9.2).

(C6)

$$B(x) = (k + \kappa(x))Q \begin{pmatrix} \cos(\theta + \upsilon(x)) & -\sin(\theta + \upsilon(x)) \\ \sin(\theta + \upsilon(x)) & \cos(\theta + \upsilon(x)) \end{pmatrix} Q^{-1} - \begin{pmatrix} 1 & 0 \\ 0 & 1 \end{pmatrix},$$

$(\mathcal{I} + B(x))x \in G$ for all $x \in G$;

(C7) $\{f, \kappa, \nu\} \subset C^{(1)}(G), \{\tau_i, i = 1, 2, \ldots, p\} \subset C^{(2)}(G)$;

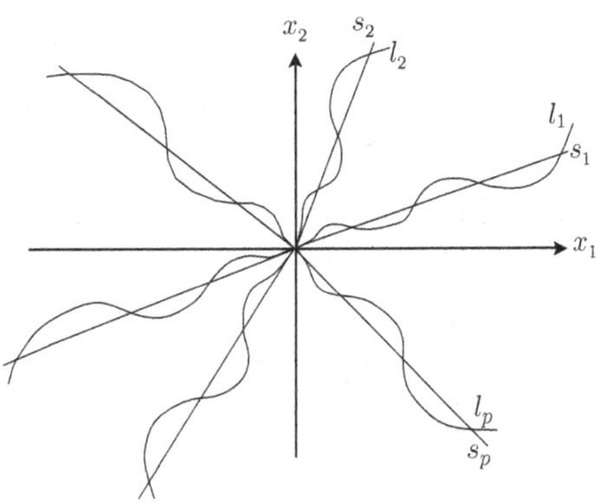

Fig. 9.2 The domain of the perturbed system (9.7) near a vertex which unites the curves l_i associated with the straight lines $s_i, i = 1, 2, \ldots, p$

(C8) $f(x) = o(||x||)$, $\kappa(x) = o(||x||)$, $\upsilon(x) = o(||x||)$, $\tau_i(x) = o(||x||^2)$, $i = 1, 2, \ldots, p$;
Moreover, we assume that the matrices A, Q, the vectors $a^i, i = 1, 2, \ldots, p$, and constants k, θ are the same as in (9.1), i.e.,

(C9) the associated with (9.7) system (9.1) is D_0-system.

If conditions (C1)–(C9) hold, then the system (9.7) is said to be a D-system. If G is sufficiently small, then conditions (C4) and (C8) imply that none of curves l_i intersect itself, they do not intersect each other, and the origin is a unique singular point of the D-system.

Exercise 9.2.1. Using the results of the last chapter, and Example 9.1.2, prove that D-system defines a B-smooth discontinuous flow.

Assume, without loss of generality, that $\gamma_i \neq \frac{\pi}{2} j$, $j = 1, 3$, and transform the equations in (C5) to the polar coordinates so that $l_i : a_i^1 r\cos(\phi) + a_i^2 r\sin(\phi) + \tau_i(r\cos(\phi), r\sin(\phi)) = 0$ or

$$\phi = \tan^{-1}(\tan\gamma_i - \frac{\tau_i}{a_i^2 r\cos(\phi)}).$$

Now, use Taylor's expansion to get that

$$l_i : \phi = \gamma_i + r\psi_i(r, \phi), \tag{9.8}$$

$i = 1, 2, \ldots, p$, where ψ_i are 2π-perodic in ϕ, continuously differentiable functions, and $\psi_i = O(r)$. If the point $x(t)$ meets the discontinuity curve l_i with an angle θ, then the point $x(\theta+)$ belongs to the curve $l_i' = \{z \in \mathbb{R}^2 | z = (\mathcal{I} + B(x))x, x \in l_i\}$. The following assertion is very important for the rest of the chapter.

Lemma 9.2.1. *Suppose (C7) and (C8) are satisfied. Then the curve l_i', $1 \leq i \leq p$, is placed between l_i and l_{i+1}, if G is sufficiently small.*

Proof. Fix $i = 1, 2, \ldots, p$, and assume that $s_i, s_{i+1}, l_i, l_{i+1}$ are transformed by the map $y = Q^{-1}x$ into lines $s_i'', s_{i+1}'', l_i'', l_{i+1}''$ respectively. Set $L_i = \{z \in \mathbb{R}^2 | z = Q^{-1}(I + B(Qy))Qy, y \in l_i''\}$, $\xi_i = Q^{-1}(I + B_0)Qs_i''$, and let $\gamma_i', \gamma_{i+1}', \zeta_i'$ be the angles of straight lines s_i'', s_{i+1}'', ξ_i. We may assume, without loss of generality, that $\gamma_i' < \zeta_i' < \gamma_{i+1}'$. To prove the lemma, it is sufficient to check whether L_i lies between curves l_i'', l_{i+1}''. Suppose that $0 < \gamma_i' < \zeta_i' < \gamma_{i+1}' < \frac{\pi}{2}$. Otherwise one can use a linear transformation, which does not change the relation of the curves. Let $c_1 y_1 + c_2 y_2 + l^*(y_1, y_2) = 0$ be the equation of the line l_i''. Use the polar coordinates $y_1 = \rho\cos(\phi)$, $y_2 = \rho\sin(\phi)$, and obtain $\phi = \gamma_i' + \rho\psi^*(\rho, \phi)$, where $\psi^*(\rho, \phi) = O(\rho)$ and ψ^* is a 2π-periodic function. If $y = (y_1, y_2) \in l_i''$ then the point

$$y^+ = Q^{-1}(B(Qy) + I)Qy, \tag{9.9}$$

where $y^+ = (y_1^+, y_2^+)$, belongs to L_i. Assume without loss of generality that $y_1^+ \neq 0$. Otherwise use the condition $y_2^+ \neq 0$. If we set $\rho = (y_1^2 + y_2^2)^{\frac{1}{2}}$, $\phi = \tan^{-1}(\frac{y_2}{y_1}), \rho^+ = ((y_1^+)^2 + (y_2^+)^2)^{\frac{1}{2}}, \phi^+ = \tan^{-1}(\frac{y_2^+}{y_1^+})$ then (9.9) implies that

$$\rho^+ = k_i\rho + \rho\beta^*(\rho, \phi), \tag{9.10}$$

$$\phi^+ = \phi + \theta + \gamma^*(\rho, \phi), \tag{9.11}$$

where β^* and γ^* are 2π-periodic in ϕ functions and $\beta^* = O(\rho), \gamma^* = O(\rho)$. Let $\sigma(y_1, y_2) = c_1y_1 + c_2y_2 + l^*(y_1, y_2)$. Then

$$\sigma(y_1^+, y_2^+) = \rho^+(c_1\cos(\phi^+) + c_2\sin(\phi^+) + l^*(\rho^+\cos(\phi^+), \rho^+\sin(\phi^+)) =$$

$$\rho^+\sqrt{c_1^2 + c_2^2}\sin(\theta + \upsilon(\rho, \phi) - \rho\psi^*(\rho, \psi)) + l^*(\rho^+\cos(\phi^+), \rho^+\sin(\phi^+)),$$

where $\upsilon(\rho, \phi) = \upsilon(Qy)$. It is readily seen that the sign of $\sigma(\rho^+, \phi^+)$ is the same as of $\sin(\theta)$, if ρ is sufficiently small. Consequently, $\sigma(\rho^+, \phi^+) > 0$. Thus, the curve L_i is placed above the curve l_i'' in the first quarter of the plane Ox_1x_2. Similarly, one can show that it is placed below l_{i+1}''. The lemma is proved. □

The last lemma guarantees that, if G is sufficiently small, then every nontrivial trajectory of the system (9.7) meets each of the lines $l_i, i = 1, 2, \ldots, p$, precisely once within any time interval of length T.

9.3 Foci of the D-System

Utilize the polar coordinates $x_1 = r\cos(\phi),\ x_2 = r\sin(\phi)$ to reduce the differential part of (9.7) to the following form:

$$\frac{dr}{d\phi} = \lambda r + P(r, \phi).$$

It is known [38, 59, 107, 117], that $P(r, \phi)$ is 2π-periodic, continuously differentiable function, and $P = o(r)$. Set $x^+ = (x_1^+, x_2^+) = (\mathcal{I} + B(x))x$, $x^+ = r^+(\cos\phi^+, \sin\phi^+)$, $\tilde{x}^+ = (\tilde{x}_1^+, \tilde{x}_2^+) = (\mathcal{I} + B(0))x$, where $x = (x_1, x_2) \in l_i, i = 1, 2, \ldots, p$. One can find that the inequality $||x^+ - \tilde{x}^+|| \leq ||B(x) - B(0)||||x||$ implies $r^+ = r + k_ir + \omega(r, \phi)$. Use the relation between $\frac{x_2^+}{x_1^+}$ and $\frac{\tilde{x}_2^+}{\tilde{x}_1^+}$ and condition (C5) to obtain that $\phi^+ = \phi + \theta_i + \gamma(r, \phi)$.

Functions ω, γ are 2π-periodic in ϕ and $\omega = o(r), \gamma(r,\phi) = o(r)$. Finally, (9.7) has the form

$$\frac{dr}{d\phi} = \lambda r + P(r,\phi),$$
$$\Delta r \mid_{(\rho,\phi)\in l_i} = k_i r + \omega(r,\phi),$$
$$\Delta \phi \mid_{(\rho,\phi)\in l_i} = \theta_i + \gamma(r,\phi). \tag{9.12}$$

It is convenient to introduce the following version of B-equivalence.

Introduce the following system:

$$\frac{d\rho}{d\phi} = \lambda \rho + P(\rho,\phi),$$
$$\Delta \rho \mid_{\phi=\gamma_i} = k_i \rho + w_i(\rho),$$
$$\Delta \phi \mid_{\phi=\gamma_i} = \theta_i, \tag{9.13}$$

where all elements, except w_i, $i = 1, 2, \ldots, p$, are the same as in (9.12) and the domain of (9.13) is $[0, 2\pi]_\phi$. Functions w_i will be defined below.

Let $r(\phi, r_0), r(0, r_0) = r_0$, be a solution of (9.12) and ϕ_i be the angle where the solution intersects l_i. Denote by $\chi_i = \phi_i + \theta_i + \gamma(r(\phi_i, r_0), \phi_i)$ the angle of $r(\phi, r_0)$ after the jump.

We shall say that systems (9.12) and (9.13) are B-equivalent in G if there exists a neighborhood $G_1 \subset G$ of the origin such that for every solution $r(\phi, r_0)$ of (9.12) whose trajectory is in G_1 there exists a solution $\rho(\phi, r_0), \rho(0, r_0) = r_0$, of (9.13) which satisfies the relation

$$r(\phi, r_0) = \rho(\phi, r_0), \phi \in [0, 2\pi]_\phi \setminus \cup_{i=1}^{p} \{[\phi_i, \hat{\gamma}_i,] \cup [\zeta_i, \hat{\chi}_i]\}, \tag{9.14}$$

and, conversely, for every solution $\rho(\phi, r_0)$ of (9.13) whose trajectory is in G_1 there exists a solution $r(\phi, r_0)$ of (9.12) which satisfies (9.14).

We will define functions w_i such that systems (9.12) and (9.13) are B-equivalent in G, if the domain is sufficiently small.

Fix i. Let $r_1(\phi, \gamma_i, \rho), r_1(\gamma_i, \gamma_i, \rho) = \rho$, be a solution of the equation

$$\frac{dr}{d\phi} = \lambda r + P(r,\phi) \tag{9.15}$$

and $\phi = \eta_i$ be the meeting angle of $r_1(\phi, \gamma_i, \rho)$ with l_i. Then

$$r_1(\eta_i, \gamma_i, \rho) = \exp(\lambda(\eta_i - \gamma_i)\rho + \int_{\gamma_i}^{\eta_i} \exp(\lambda(\eta_i - s)P(r_1(s, \gamma_i, \rho), s)ds.$$

Let $\eta_i^1 = \eta_i + \theta_i + \gamma(r_1(\eta_i, \gamma_i, \rho), \eta_i), \rho^1 = (1 + k_i) r_1(\eta_i, \gamma_i, \rho) + \omega(r(\eta_i, \gamma_i, \rho), \eta_i)$, and $r_2(\phi, \eta_i^1, \rho^1)$ be the solution of system (9.15),

$$r_2(\zeta_i, \eta_i^1, \rho^1) = \exp(\lambda(\zeta_i - \eta_i^1))\rho^1 + \int_{\eta_i^1}^{\zeta_i} \exp(\lambda(\zeta_i - s) P(r_2(s, \eta_i^1, \rho^1), s) ds.$$

Introduce

$$w_i(\rho) = r_2(\zeta_i, \eta_i^1, \rho^1) - (1 + k_i)\rho = \exp(\lambda(\zeta_i - \eta_i^1)[(1 + k_i)(\exp(\lambda(\eta_i - \gamma_i))\rho +$$
$$\int_{\gamma_i}^{\eta_i} \exp(\lambda(\eta_i - s)) P(r_1(s, \gamma_i, \rho), s) ds) + \omega(r_1(\eta_i, \gamma_i, \rho), \eta_i)] +$$
$$\int_{\eta_i^1}^{\zeta_i} \exp(\lambda(\zeta_i - s) P(r_2(s, \eta_i^1, \rho^1), s) ds - (1 + k)\rho$$

or, if simplified,

$$\begin{aligned} &w_i(\rho) = (1 + k)[\exp(-\lambda\gamma(r_1(\eta_i, \gamma_i, \rho), \eta_i)) - 1]\rho + \\ &(1 + k) \int_{\gamma_i}^{\eta_i} \exp(\lambda(\zeta_i - \theta_i - s - \rho\gamma(r_1(\eta_i, \gamma_i, \rho), \eta_i))) P(r_1(s, \gamma_i, \rho), s) ds + \\ &\int_{\eta_i^1}^{\zeta_i} \exp(\lambda(\zeta_i - s)) P(r_2(s, \eta_i^1, \rho^1), s) ds + \\ &\exp(\lambda(\zeta_i - \eta_i^1))\omega(r_1(\eta_i, \gamma_i, \rho), \eta_i). \end{aligned} \tag{9.16}$$

Differentiating (9.8) and (9.16) one can find that

$$\frac{d\eta_i}{d\rho} = \frac{\frac{\partial r_1}{\partial \rho}[\psi_i + r_1 \frac{\partial \psi_i}{\partial r}]}{1 - (\lambda r_1 + P)[\psi_i + r_1 \frac{\partial \psi_i}{\partial r}] - r_1 \frac{\partial \psi_i}{\partial \phi}}, \quad \frac{d\eta_i^1}{d\rho} = \frac{d\eta_i}{d\rho}(1 + \frac{\partial \gamma}{\partial \phi}) + \frac{\partial \gamma}{\partial r}\frac{\partial r_1}{\partial \rho},$$

$$\begin{aligned} &\frac{dw_i}{d\rho} = (1 + k_i)[\mathrm{e}^{-\lambda\gamma} - 1] - \lambda(1 + k_i)\mathrm{e}^{-\lambda\gamma}(\frac{\partial \gamma}{\partial r}\frac{\partial r_1}{\partial \rho} + \frac{\partial \gamma}{\partial \phi}\frac{d\eta_i}{d\rho})\rho + \\ &(1 + k_i)\mathrm{e}^{\lambda(\zeta_i - \theta_i - \eta_i - \gamma)} P \frac{d\eta_i}{d\rho} + \\ &(1 + k_i) \int_{\gamma_i}^{\eta_i} \mathrm{e}^{\lambda(\zeta_i - \theta - s - \gamma)} \{-\lambda(\frac{\partial \gamma}{\partial r}\frac{\partial r_1}{\partial \rho} + \frac{\partial \gamma}{\partial \phi}\frac{d\eta_i}{d\rho}) P - \frac{\partial P}{\partial r}\frac{\partial r_1}{\partial \rho} - \frac{\partial P}{\partial \phi}\frac{d\eta_i}{d\rho}\} ds + \\ &\int_{\eta_i^1}^{\zeta_i} \mathrm{e}^{\lambda(\zeta_i - s)} \frac{\partial P(r_2(s, \eta_i^1, \rho^1), s)}{\partial r}\frac{\partial r_2}{\partial \rho} ds - \mathrm{e}^{\lambda(\zeta_i - \eta_i^1)} P(\rho^1, \eta_i^1)\frac{\partial \eta_i^1}{\partial \rho} + \\ &\mathrm{e}^{\lambda(\zeta_i - \eta_i^1)}[-\frac{\partial \eta_i^1}{\partial \rho}\omega + \frac{\partial \omega}{\partial r}\frac{\partial r_1}{\partial \rho} + \frac{\partial \omega}{\partial \phi}\frac{d\eta_i}{d\rho}]. \end{aligned} \tag{9.17}$$

Analyzing (9.16) and (9.17) one can prove that the following two lemmas are valid.

Lemma 9.3.1. *If conditions (C1)–(C5) are valid then w_i is a continuously differentable function, and $w_i(\rho) = o(\rho), i = 1, 2, \ldots, p$.*

Lemma 9.3.2. *The systems (9.12) and (9.13) are B-equivalent if G is sufficiently small.*

Theorem 9.3.1. *Suppose that (C1)–(C6) are satisfied and $q < 1$ ($q > 1$). Then the origin is a stable (unstable) focus of system (9.7).*

Proof. Let $r(\phi, r_0), r(0, r_0) = r_0$, be the solution of (9.12), and $\rho(\phi, r_0)$, $\rho(0, r_0) = r_0$, be the solution of (9.13). Using ψ-substitution one can obtain that

$$
\begin{aligned}
&\rho(\phi, r_0) = \exp(\lambda\phi)\{\Pi_{i=1}^{m}(1 + k_i)\exp(-\lambda\sum_{s=1}^{m}\theta_s)r_0 + \\
&\Pi_{i=1}^{m}(1 + k_i)\exp(-\lambda\sum_{s=1}^{m}\theta_s)\int_0^{\gamma_1}\exp(-\lambda u)P du + \\
&\Pi_{i=2}^{m}(1 + k_i)\exp(-\lambda\sum_{s=2}^{m}\theta_s)\int_{\zeta_1}^{\gamma_2}\exp(-\lambda u)P du + \ldots \\
&\int_{\zeta_m}^{\phi}\exp(-\lambda u)P du + \Pi_{i=2}^{m}(1 + k_i)\exp(-\lambda\sum_{s=2}^{m}\theta_s)w_1 + \\
&\Pi_{i=3}^{m}(1 + k_i)\exp(-\lambda\sum_{s=3}^{m}\theta_s)w_2 \ldots + \exp(-\lambda\zeta_m)w_m\},
\end{aligned} \tag{9.18}
$$

where $\phi \in [0, 2\pi]_\phi, P = P(\rho(\phi, r_0), \phi), w_i = w_i(\rho(\gamma_i, r_0))$. Now, applying Theorem 6.1.1, conditions (C4), (C5), and Lemma 9.3.1 one can find that the solution $\rho(\psi, r_0)$ is differentiable in r_0 and the derivative $\frac{\partial\rho(\phi, r_0)}{\partial r_0}$ at the point $(2\pi, 0)$ is equal to q. Since (9.12) and (9.13) are B-equivalent it follows that:

$$
\frac{\partial r(2\pi, 0)}{\partial r_0} = q
$$

and the proof is completed. □

9.4 The Center and Focus Problem

Throughout this section we assume that $q = 1$. That is, the critical case is considered. Functions $f, \kappa, v, \tau_i, i = 1, 2, \ldots, p$, are assumed to be analytic in G. By condition (C8), Taylor's expansions of functions f, κ, and v start with members of order not less than 2, and the expansions of $\tau_i, i = 1, 2, \ldots, p$, start with members of order not less than 3. First, we investigate the problem for (9.13) all of whose elements are analytic functions, if ρ is sufficiently small. Theorem 6.4.2 implies that $w_i, i = 1, 2, \ldots, p$, are analytic functions in ρ and the solution $\rho(\phi, r_0)$ of equation (9.13) has the following expansion:

$$\rho(\phi, r_0) = \sum_{i=0}^{\infty} \rho_i(\phi) r_0^i, \tag{9.19}$$

where $\phi \notin (\gamma_i, \zeta_i], i = 1, 2, \ldots, p, \rho_0(\phi) = 0, q = \rho_1(\phi) = 1$. One can define the Poincaré return map

$$\rho(2\pi, r_0) = \sum_{i=1}^{\infty} a_i r_0^i, \tag{9.20}$$

where $a_i = \rho_i(2\pi), i \geq 1, a_1 = q = 1$. The expansions exist, see Sect. 6.4, such that

$$P(\rho, \phi) = \sum_{i=2}^{\infty} P_i(\phi) \rho^i,$$
$$w_j(\rho) = \sum_{i=2}^{\infty} w_{ji} \rho^i, \tag{9.21}$$

where $P_i(\phi), w_{ji}(\phi), j \geq 2$, are 2π-periodic functions which can be defined by using (9.12). The coefficient $\rho_j(\phi), j \geq 2$, is the solution of the system

$$\frac{d\rho}{d\phi} = P_j(\phi),$$
$$\Delta\rho \mid_{\phi \neq \gamma_i} = w_{ji},$$
$$\Delta\phi \mid_{\phi \neq \gamma_i} = \theta_i, \tag{9.22}$$

with the initial condition $\rho_j(0) = 0$. Hence, coefficients of (9.20) are equal to

$$a_j = \int_0^{\gamma_1} P_j(\phi) d\phi + \sum_{i=1}^{p-1} \int_{\zeta_i}^{\gamma_{i+1}} P_j(\phi) d\phi + \int_{\zeta_p}^{2\pi} P_j(\phi) d\phi + \sum_{i=1}^{p} w_{ji}. \tag{9.23}$$

From (9.20) and (9.23) it follows that the following lemma is true.

Lemma 9.4.1. *Let $q = 1$ and the first nonzero element of the sequence $a_j, j \geq 2$, be negative (positive), then the origin is a stable (unstable) focus of (9.13). If $a_j = 0, j \geq 2$, then the origin is a center of (9.13).*

B-equivalence of systems (9.12) and (9.13) implies immediately that the following theorem is valid.

Theorem 9.4.1. *Let $q = 1$ and the first nonzero element of the sequence a_j, $j \geq 2$, be negative (positive), then the origin is a stable (unstable) focus of (9.7). If $a_j = 0$ for all $j \geq 2$, then the origin is a center of (9.7).*

9.5 Bifurcation of a Discontinuous Limit Cycle

We consider the following system:

$$\begin{aligned} \frac{dx}{dt} &= Ax + f(x) + \mu F(x,\mu), \\ \Delta x|_{x\in\Gamma(\mu)} &= B(x,\mu)x. \end{aligned} \tag{9.24}$$

To establish the Hopf bifurcation theorem we need the following assumptions:

(A1) the set $\Gamma(\mu) = \cup_{i=1}^{p} l_i(\mu)$ is a union of curves in G, which start at the origin and do not include it, $l_i : (a^i, x) + \tau_i(x) + \mu\nu(x,\mu) = 0, 1 \leq i \leq p$;
(A2) there exist a matrix $Q(\mu) \in \mathcal{R}$, $Q(0) = Q$, analytic in $(-\mu_0, \mu_0)$, and real numbers γ, χ such that $Q^{-1}(\mu)B(x,\mu)Q(\mu) =$

$$(k+\mu\gamma+\kappa(x))\begin{pmatrix} \cos(\theta+\mu\chi+\upsilon(x)) & -\sin(\theta+\mu\chi+\upsilon(x)) \\ \sin(\theta+\mu\chi+\upsilon(x)) & \cos(\theta+\mu\chi+\upsilon(x)) \end{pmatrix} - \begin{pmatrix} 1 & 0 \\ 0 & 1 \end{pmatrix};$$

(A3) associated with (9.24) systems

$$\begin{aligned} \frac{dx}{dt} &= Ax, \\ \Delta x|_{x\in\Gamma(0)} &= B_0 x, \end{aligned} \tag{9.25}$$

and

$$\begin{aligned} \frac{dx}{dt} &= Ax + f(x), \\ \Delta x|_{x\in\Gamma(0)} &= B(x,0)x, \end{aligned} \tag{9.26}$$

are D_0-system and D-system, respectively;
(A4) functions $\kappa, \upsilon : G \to \mathbb{R}^2$ and $F, \nu : G \times (-\mu_0, \mu_0) \to \mathbb{R}^2$ are analytic in $G \times (-\mu_0, \mu_0)$;
(A5) $F(0,\mu) = 0, \nu(0,\mu) = 0$, for all $\mu \in (-\mu_0, \mu_0)$.

Additionally, we shall need the following system:

$$\begin{aligned} \frac{dx}{dt} &= A(\mu)x, \\ \Delta x|_{x\in\Gamma_0(\mu)} &= B(0,\mu)x, \end{aligned} \tag{9.27}$$

where $A(\mu) = A + \mu\frac{\partial F(0,\mu)}{\partial x}$, and $\Gamma_0(\mu) = \cup_{i=1}^{p} m_i$ with

$$m_i: \quad (a^i + \mu\frac{\partial v(0,\mu)}{\partial x}, x) = 0, \quad i = 1, 2, \ldots, p.$$

The polar transformation takes (9.24) to the following form:

$$\begin{aligned} &\frac{dr}{d\phi} = \lambda r + P(r, \phi, \mu), \\ &\Delta r \mid_{(r,\phi)\in l_i(\mu)} = k_i r + \omega(r, \phi, \mu), \\ &\Delta \phi \mid_{(r,\phi)\in l_i(\mu)} = \theta_i + r\gamma(r, \phi, \mu). \end{aligned} \tag{9.28}$$

The functions $w_i(\rho, \mu)$ can be defined in the same manner as in (9.16) such that the system

$$\begin{aligned} &\frac{d\rho}{d\phi} = \lambda\rho + P(\rho, \phi, \mu), \ \phi \neq \gamma_i(\mu), \\ &\Delta\rho \mid_{\phi=\gamma_i(\mu)} = k_i\rho + w_i(\rho, \mu), \\ &\Delta\phi \mid_{\phi=\gamma_i(\mu)} = \theta_i(\mu), \end{aligned} \tag{9.29}$$

where $\gamma_i(\mu), i = 1, 2, \ldots, p$, are angles of m_i, is B-equivalent to (9.28).

Similarly to (9.6) one can define the function

$$q(\mu) = \exp(\lambda(\mu)(2\pi - \sum_{j=1}^{p}(\zeta_j(\mu) - \gamma_j(\mu))\Pi_{j=p}^{1}(1 + k_j(\mu)) \tag{9.30}$$

for system (9.27). Theorem 6.4.2 of Chap. 6 implies that $q(\mu)$ is an analytic function.

Theorem 9.5.1. *Assume that $q(0) = 1$, $q'(0) \neq 0$ and the origin is a focus of (9.26). Then, for sufficiently small r_0, there exists a continuous function $\mu = \delta(r_0), \delta(0) = 0$, such that the solution $r(\phi, r_0, \delta(r_0))$ of (9.28) is periodic function with period 2π. The period of the corresponding solution of (9.24) is $T = (2\pi - \sum_{i=1}^{p}\theta_i)\beta^{-1} + o(|\mu|)$. Moreover, if the origin is a stable focus of (9.26) then the closed trajectory is a limit cycle.*

Proof. If $\rho(\phi, r_0, \mu)$ is a solution of (9.29), then by Theorem 6.4.2 we have that

$$\rho(2\pi, r_0, \mu) = \sum_{i=1}^{\infty} a_i(\mu) r_0^i,$$

where $a_i(\mu) = \sum_{j=0}^{\infty} a_{ij}\mu^j, a_{10} = q(0) = 1, a_{11} = q'(0) \neq 0$. Define the displacement function

$$\mathcal{V}(r_0, \mu) = \rho(2\pi, r_0, \mu) - r_0 = q'(0)\mu r_0 + \sum_{i=2}^{\infty} a_{i0} r_0^i + r_0\mu^2 G_1(r_0, \mu) + r_0^2\mu G_2(r_0, \mu),$$

where G_1, G_2 are functions analytic in a neighborhood of $(0,0)$. The bifurcation equation is $\mathcal{V}(r_0,\mu) = 0$. Canceling by r_0 one can rewrite the equation as

$$\mathcal{H}(r_0,\mu) = 0, \tag{9.31}$$

where

$$\mathcal{H}(r_0,\mu) = q'(0)\mu + \sum_{i=2}^{\infty} a_{i0} r_0^{i-1} + \mu^2 G_1(r_0,\mu) + r_0\mu G_2(r_0,\mu)$$

Since

$$\mathcal{H}(0,0) = 0, \qquad \frac{\partial \mathcal{H}(0,0)}{\partial \mu} = q'(0) \neq 0,$$

for sufficiently small r_0 there exists a function $\mu = \delta(r_0)$ such that $r(\phi, r_0, \delta(r_0))$ is a periodic solution. If conditions $a_{i0} = 0, i = 2,\ldots,l-1$, and $a_{l0} \neq 0$ are valid, then one can obtain from (9.31) that

$$\delta(r_0) = -\frac{a_{l0}}{q'(0)} r_0^{l-1} + \sum_{i=l}^{\infty} \delta_i r_0^i. \tag{9.32}$$

By analysis of the latter expression one can conclude that the bifurcation of periodic solutions emerges if the focus is stable with $\mu = 0$ and unstable with $\mu \neq 0$ and conversely. If $\rho(\phi) = \rho(\phi, \bar{r_0}, \bar{\mu})$ is a periodic solution of (9.29), then it is known that the trajectory is a limit cycle if

$$\frac{\partial \mathcal{V}(\bar{r_0},\bar{\mu})}{\partial r_0} < 0. \tag{9.33}$$

We have that

$$\frac{\partial \mathcal{V}(r_0,\mu)}{\partial r_0} = q'(0)\mu + \sum_{i=2}^{\infty} i a_{i0} r_0^{i-1} + \mu^2 G_1(r_0,\mu) + 2r_0\mu G_2(r_0,\mu).$$

Let a_{l0} be the first nonzero element among a_{i0} and $a_{l0} < 0$. Using (9.32), one can obtain that

$$\frac{\partial \mathcal{V}(\bar{r_0},\bar{\mu})}{\partial r_0} = (l-1) a_{l0} \bar{r}_0^{l-1} + Q(\bar{r_0}),$$

where Q starts with a member whose order is not less than l. Hence, (9.33) is valid. Now, B-equivalence of (9.28) and (9.29) proves the theorem. □

Remark 9.5.1. (a) It is important to notice that the bifurcation theorem can be obtained by applying the results in [83] and theorems of Chap. 6. We follow the approach which is focused on the expansions of solutions [107].

(b) To illustrate that discontinuous dynamical systems may provide more interesting opportunities than continuous dynamics, let us compare the bifurcation diagrams of an ordinary differential equation, Fig. 9.3, and a discontinuous dynamical system of type (9.24), Fig. 9.4. One can see that the first diagram resembles a bud, and the second one a rose. They demonstrate that a theory of differential equations flourishes if a discontinuity is involved in analysis.

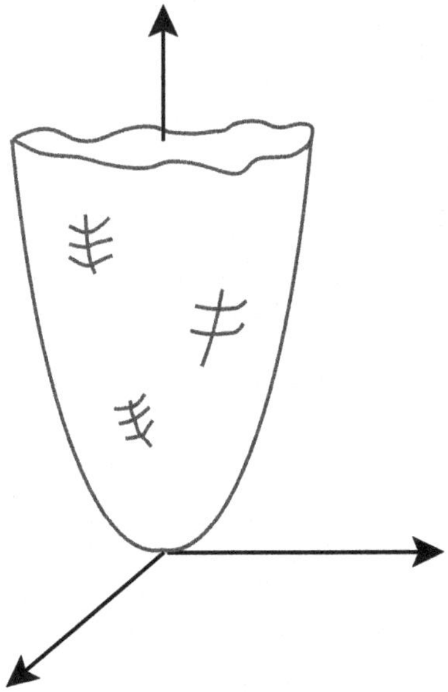

Fig. 9.3 A Hopf bifurcation diagram of an ordinary differential equation

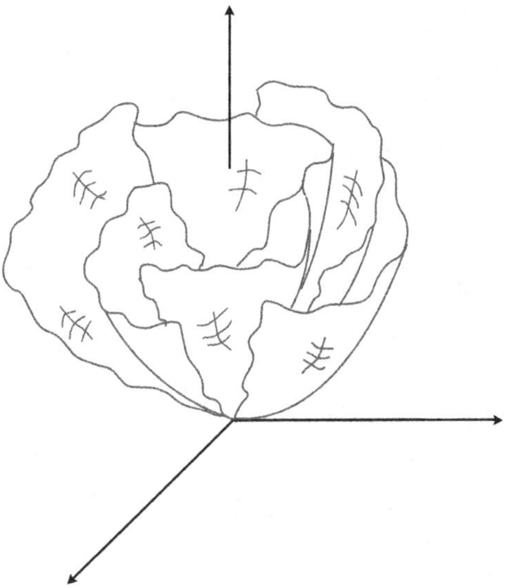

Fig. 9.4 A Hopf bifurcation diagram of a discontinuous dynamical system

9.6 Examples

Example 9.6.1. Consider the following system:

$$
\begin{aligned}
x_1' &= (2+\mu)x_1 - x_2 + x_1^2 x_2,\\
x_2' &= x_1 + (2+\mu)x_2 + 3x_1^3 x_2,\\
\Delta x_1|_{x\in l} &= ((\kappa+\mu^2)\cos(\frac{\pi}{6}) - 1)x_1 - (\kappa+\mu^2)\sin(\frac{\pi}{6})x_2,\\
\Delta x_2|_{x\in l} &= (\kappa+\mu^2)\sin(\frac{\pi}{6})x_1 + ((\kappa+\mu^2)\cos(\frac{\pi}{6}) - 1)x_2,
\end{aligned}
\tag{9.34}
$$

where $\kappa = e^{-\frac{11\pi}{6}}$, and the curve l is given by the equation $x_2 = x_1^3$, where $x_1 > 0$. One can define, using (9.30), that $q(\mu) = (\kappa+\mu^2)\exp((2+\mu)\frac{11\pi}{6}), q(0) = \kappa\exp(\frac{11\pi}{3}) = 1, q'(0) = -\frac{11\pi}{6} \neq 0$. Thus, by Theorem 9.5.1, system (9.34) has a periodic solution with period $\approx \frac{11\pi}{12}$ if $|\mu|$ is sufficiently small.

Example 9.6.2. Let the following system be given:

$$
\begin{aligned}
x_1' &= (\mu-1)x_1 - x_2,\ x_2' = x_1 + (\mu-1)x_2,\\
\Delta x_1|_{x\in l} &= ((\kappa - x_1^2 - x_2^2)\cos(\frac{\pi}{4}) - 1)x_1 - (\kappa - x_1^2 - x_2^2)\sin(\frac{\pi}{4})x_2,\\
\Delta x_2|_{x\in l} &= (\kappa - x_1^2 - x_2^2)\sin(\frac{\pi}{4})x_1 + ((\kappa - x_1^2 - x_2^2)\cos(\frac{\pi}{4}) - 1)x_2,
\end{aligned}
\tag{9.35}
$$

where l is a curve given by the equation $x_2 = x_1 + \mu x_1^2, x_1 > 0, \kappa = \exp(\frac{7\pi}{4})$. Using (9.30) one can find that $q(\mu) = \kappa\exp((\mu-1)\frac{7\pi}{4}), q(0) = \kappa\exp(-\frac{7\pi}{4}) = 1$, $q'(0) = \frac{7\pi}{4} \neq 0$. Moreover, one can see that for the associated D-system

$$
\begin{aligned}
x_1' &= -x_1 - x_2,\ x_2' = x_1 - x_2,\\
\Delta x_1|_{x\in s} &= ((\kappa - x_1^2 - x_2^2)\cos(\frac{\pi}{4}) - 1)x_1 - (\kappa - x_1^2 - x_2^2)\sin(\frac{\pi}{4})x_2,\\
\Delta x_2|_{x\in s} &= (\kappa - x_1^2 - x_2^2)\sin(\frac{\pi}{4})x_1 + ((\kappa - x_1^2 - x_2^2)\cos(\frac{\pi}{4}) - 1)x_2,
\end{aligned}
\tag{9.36}
$$

where s is given by the equation $x_2 = x_1, x_1 > 0$, the origin is a stable focus. Indeed, using polar coordinates, denote by $r(\phi, r_0)$ the solution of (9.36) starting at the angle $\phi = \frac{\pi}{4}$. We can define that $r(\frac{\pi}{4} + 2\pi n, r_0) = (\kappa - r^2(\frac{\pi}{4} + 2\pi(n-1), r_0))\exp(-\frac{7\pi}{4})$. From the last expression it is easily seen that the sequence $r_n = r(\frac{\pi}{4} + 2\pi n, r_0)$ is monotonically decreasing and there exists a limit of r_n. Assume that $r_n \to \sigma \neq 0$. Then it implies that there exists a periodic solution of (9.36) and $\sigma = (\kappa - \sigma^2)\exp(-\frac{7\pi}{4})\sigma$ which is a contradiction. Thus, $\sigma = 0$. Consequently, the origin is a stable focus of (9.36) and by Theorem 9.5.1 the system (9.35) has a limit cycle with period $\approx \frac{7\pi}{4}$ if $\mu > 0$ is sufficiently small.

Notes

The present chapter contains mainly results of paper [4], and is based on the perturbation theory, which was founded by H. Poincaré and A. M. Lyapunov [104, 132], and the bifurcation methods [38, 48, 71, 80, 83, 106, 107, 134, 157]. The main result is the bifurcation of a periodic solution from the equilibrium of the discontinuous dynamical system. After the initial impetus of H. Poincaré [132], A. Andronov [38], and E. Hopf [80] this method of research of periodic motions has been used very successfully for various differential equations by many authors (see [66, 71, 83, 107] and references cited there). There have been two principal obstacles of expansion of this method for discontinuous dynamical systems. While the absence of developed differentiability of solutions has been the first one, the choice of a nonperturbed system convenient to study has been the second. The present investigation utilizes extensively the differentiability and analyticity of discontinuous solutions discussed in Chap. 6. The nonperturbed equation is specifically defined. The results of the present chapter can be extended by the dimension enlarging [21] and application to differential equations with discontinuous right side [13]. They are applied to control the population dynamics [14], and can be effectively employed in mechanics, electronics, biology, and medicine [38, 52, 71, 107, 115, 123].

Chapter 10
Chaos and Shadowing

10.1 Introduction and Preliminaries

The proof of the existence of chaotic attractors remains an important and difficult problem, which is still not resolved fully, even for the Lorentz system [49, 72, 84, 150]. In this chapter, a multidimensional chaos is generated by a special initial value problem for the nonautonomous impulsive differential equation. The existence of a chaotic attractor is shown, where density of periodic solutions, sensitivity of solutions, and existence of a trajectory, which is dense in the set of all orbits are observed. That is, we concentrate on the topological ingredients of the version proposed by Devaney [62]. An appropriate example is constructed, where a chaotic attractor is indicated, and the intermittency is observed.

The discontinuous system consists of an impulsive differential equation and of a discrete equation, which generates the moments of impacts.

We suppose that the generator is chaotic while the impulsive system is dissipative for all possible sequences of moments of discontinuities, and we prove that the system has a similar chaotic nature. Similarly, if the generator function has a shadowing property [40,55,76,134], then the system admits an analogue of the property. The shadowing exists if the generator is uniformly hyperbolic on the invariant set of initial moments, or a nonhyperbolic map.

The results of this chapter illustrate that impulsive differential equations may play a special role in the investigation of the complex behavior of dynamical systems.

Finally, one must say that the B-equivalence method is used to obtain main results of this chapter. Thus, we will complete the integrity of the book.

Let us consider a continuous map $H : I \to \mathbb{R}, I = [0, 1]$, with a positively invariant compact set $\Lambda \subseteq I$. Let $\kappa_{i+1} = H(\kappa_i), \kappa_0 = t_0 \in \Lambda$, and the sequence $\zeta(t_0) = \{\zeta_i(t_0)\}$ be defined, where $\zeta_i(t_0) = i + \kappa_i(t_0), i \geq 0$.

One may consider the logistic map $h(t, \mu) = \mu t(1 - t), \mu > 0$, as an example of H. The main object of discussion in this chapter is the following special initial value problem,

$$z'(t) = Az(t) + f(z),$$

M. Akhmet, *Principles of Discontinuous Dynamical Systems*,
DOI 10.1007/978-1-4419-6581-3_10,

$$\begin{aligned}\Delta|_{t=\zeta_i(t_0)} &= Bz(\zeta_i(t_0)) + W(z(\zeta_i(t_0))),\\ z(t_0) &= z_0,\ (t_0, z_0) \in \Lambda \times \mathbb{R}^n,\end{aligned} \tag{10.1}$$

where $z \in \mathbb{R}^n, t_0 \in I, t \geq t_0$.

We shall need the following basic assumptions for the problem:

(C1) A, B are $n \times n$ constant real valued matrices; $\det(\mathcal{I} + B) \neq 0$, where $\mathcal{I}$ is the identical matrix;

(C2) for all $x_1, x_2 \in \mathbb{R}^n$ functions $f(x) : \mathbb{R}^n \to \mathbb{R}^n$, $W : \mathbb{R}^n \to \mathbb{R}^n$, satisfy

$$||f(x_1) - f(x_2)|| + \|W(x_1) - W(x_2)\| \leq L||x_1 - x_2||,$$

where $L > 0$ is a constant;

(C3) $\sup_{x\in\mathbb{R}^n} ||f(x)|| + \sup_{x\in\mathbb{R}^n} ||W(x)|| = M_0 < \infty$;

(C4) the matrices A and B commute and the real parts of all eigenvalues of $A + \ln(\mathcal{I} + B)$ are negative.

From the previous chapters it implies that under these conditions a solution $z(t) = z(t, t_0, z_0), z_0 \in \mathbb{R}^n$ of (10.1) exists, and is unique on $[t_0, \infty)$.

Consider an unbounded and strictly increasing sequence θ with elements θ_i, $i - 1 < \theta_i < i + 2, i \in \mathbb{Z}$. Let us denote by $Z(t, s)$ the transition matrix of the linear homogeneous system

$$\begin{aligned}z'(t) &= Az(t),\\ \Delta z|_{t=\theta_i} &= Bz(\theta_i).\end{aligned} \tag{10.2}$$

Condition $(C4)$ and the result of Exercise 4.1.8 imply that there exist two positive numbers N and ω, which do not depend on θ, such that $||Z(t, s)|| \leq Ne^{-\omega(t-s)}, t \geq s$. In what follows, we shall denote by $Z(t, s, \xi)$ the transition matrix $Z(t, s)$ if $\theta = \zeta(\xi)$.

We shall need the following additional assumptions:

(C6) $NL[\frac{2}{\omega} + \frac{e^{\omega}}{1-e^{-\omega}}] < 1$;

(C7) $-\omega + NL + \ln(1 + NL) < 0$.

The solution $z(t) = z(t, t_0, z_0)$ of (10.1) satisfies the following integral equation:

$$z(t) = Z(t, t_0, t_0)z_0 + \int_{t_0}^{t} Z(t, s, t_0) f(z(s))ds + \sum_{t_0 \leq \zeta_i < t} Z(t, \zeta_i(t_0), t_0) W(z(\zeta_i(t_0))).$$

Using the last formula and technique of Chap. 7 (see Theorem 7.1.5), one can verify that all solutions eventually, as t increases, enter the tube with the radius $M = NM_0[\frac{1}{\omega} + \frac{e^{\omega}}{1-e^{-\omega}}], t \in \mathbb{R}$. That is, the discussion of this chapter can be made assuming that all solutions are inside the tube. Moreover, if the sequence $\kappa(t_0)$ is periodic with a period $p \in \mathbb{N}$, then there is a solution of (10.1) with the same period, and its integral curve is placed in the tube.

We assume that:

(C8) $Bx + W(x) \neq 0$, if $\|x\| \leq M$.

The last condition implies that periodic solutions are different for different p.

Denote by $\mathcal{PC}$ the set of all solutions $z(t) = z(t, t_0, z_0), t_0 \in \Lambda, z_0 \in \mathbb{R}^n, t \geq t_0$ of (10.1), and denote $\mathcal{PCA} = \{z \in \mathcal{PC} : \|z(t_0)\| < M, t_0 \in \Lambda\}$. In the next section, we define conditions with which $\mathcal{PCA}$ is a chaotic attractor.

10.2 The Devaney's Chaos

Let us assume that the map H admits all Devaney's ingredients of chaos on the set Λ, that is:

1. there exists a positive δ_0 such that for each $t \in \Lambda$ and $\epsilon > 0$ there is a point $\tilde{t} \in \Lambda$ with $|t - \tilde{t}| < \epsilon$ and $|H^i(t) - H^i(\tilde{t})| \geq \delta_0$, for some positive integer i (sensitivity);
2. there exists an element $t^* \in \Lambda$ such that the set $H^i(t^*), i \geq 0$, is dense in Λ (transitivity);
3. the set of period$-p$ points, $p \geq 1$, is dense in Λ (density of periodic points).

Let us define the chaos for the discontinuous dynamics of (10.1).

Definition 10.2.1. We say that (10.1) is sensitive on Λ if there exist positive real numbers ϵ_0, ϵ_1 such that for each $t_0 \in \Lambda$, and $\delta > 0$ one can find a number $t_1 \in \Lambda, |t_0 - t_1| < \delta$, such that for each couple of solutions $z(t) = z(t, t_0, z_0), z_1(t) = z(t, t_1, z_1), z_0, z_1 \in \mathbb{R}^n$, there exists an interval $Q \subset [t_0, \infty)$ with the length not less than ϵ_1 such that $||z(t) - z_1(t)|| \geq \epsilon_0, t \in Q$, and there are no points of discontinuity of $z(t), z_1(t)$ in Q.

We shall denote $z(t)(\epsilon, J)z_1(t)$, if solutions $z(t)$ and $z_1(t)$ of (10.1), $z(t) = z(t, t_0, z_0), z_1(t) = z(t, t_1, z_1), t_0, t_1 \in \Lambda$, are ϵ-equivalent on J. The concept of the equivalence is described in Sect. 5.4.

Definition 10.2.2. The set of all periodic solutions $\phi(t) = \phi(t, t_0), t_0 \in \Lambda$, of (10.1) is called dense in $\mathcal{PC}$ if for every solution $z(t) \in \mathcal{PC}$ and each $\epsilon > 0, E > 0$, there exist a periodic solution $\phi(t, t^*), t^* \in \Lambda$, and an interval $J \subset [t_0, \infty)$ with the length E such that $\phi(t)(\epsilon, J)z(t)$.

Definition 10.2.3. A solution $z_*(t) \in \mathcal{PC}$ of (10.1) is called dense in the set of all orbits of $\mathcal{PC}$ if for every solution $z(t) \in \mathcal{PC}$ of (10.1), and each $\epsilon > 0, E > 0$, there exist an interval $J \subset [0, \infty)$ with the length E and a real number ξ such that $z_*(t + \xi)(\epsilon, J)z(t)$.

Definition 10.2.4. The problem (10.1) is chaotic if: (*i*) it is sensitive; (*ii*) the set of all periodic solutions $\phi(t, t_0), t_0 \in \Lambda$, is dense in $\mathcal{PC}$; (*iii*) there exists a solution $z_*(t)$, which is dense in $\mathcal{PC}$.

Remark 10.2.1. Definitions of the chaotic ingredients have been worked out in detail issuing from the two reasons: the considered system is nonautonomous and consequently we analyze integral curves, but not trajectories; the system is impulsive and different solutions have different points of discontinuity that necessitates the B-topology.

Theorem 10.2.1. *Assume that conditions (C1)–(C6) are fulfilled. Then the set of all periodic solutions $\phi(t,t_0), t_0 \in \Lambda$, of (10.1) is dense in $\mathcal{PC}$.*

Proof. Fix $t_1 \in \Lambda$ and $E, \epsilon > 0$. The density of periodic points of H and uniform continuity of this map imply that for an arbitrary large number $\tilde{T}$ there exists a sequence $\zeta(t_0)$, defined by a periodic sequence $\kappa(t_0)$, such that $\|\zeta(t_1) - \zeta(t_0)\|_Q < \epsilon$, where $Q = (t_1, t_1 + \tilde{T} + E)$. We shall find the number $\tilde{T}$ so large that solution $z(t) = z(t, t_1, z_1)$, $\|z_1\| < M$, is ϵ-equivalent to $\phi(t,t_0)$ on $J = (t_1 + \tilde{T}, t_1 + \tilde{T} + E)$.

Denote by $Z_1(t,s) = Z(t,s,t_1)$ and $Z_2(t,s) = Z(t,s,t_0), t \geq s$, the transition matrices. We have that

$$z(t) = Z_1(t,1)z(1) + \int_{c1}^{t} Z_1(t,s)f(z(s))ds + \sum_{1 \leq \zeta_i < t} Z_1(t,\zeta_i(t_1))W(z(\zeta_i(t_1))),$$

$$\phi(t) = Z_2(t,1)\phi(1) + \int_{c1}^{t} Z_2(t,s)f(\phi(s))ds + \sum_{1 \leq \zeta_i < t} Z_2(t,\zeta_i(t_0))W(\phi(\zeta_i(t_0))).$$

The difference between $z(t)$ and $\phi(t)$ cannot be evaluated by using the last two expressions since the moments of discontinuities do not coincide. The method of B-equivalence is helpful here. Introduce the following B-maps

$$W_i^1(z) = (\mathcal{I} + B)\Big[(\mathrm{e}^{A(\zeta_i(t_1) - \zeta_j(t_0))} - \mathcal{I})z + \int_{\zeta_j(t_0)}^{\zeta_i(t_1)} \mathrm{e}^{A(\zeta_i(t_1) - s)} f(z(s))ds\Big] +$$

$$W((\mathcal{I} + B)[\mathrm{e}^{A(\zeta_i(t_1) - \zeta_j(t_0))}z + \int_{\zeta_j(t_0)}^{\zeta_i(t_1)} \mathrm{e}^{A(\zeta_i(t_1) - s)} f(z(s))ds]) -$$

$$\int_{\zeta_j(t_0)}^{\zeta_i(t_1)} \mathrm{e}^{A(\zeta_i(t_1) - s)} f(z_1(s))ds - W(z),$$

where $z(t), z_1(t), z(\zeta_i(t_0)) = z, z_1(\zeta_i(t_1)) = z(\zeta_i(t_1)+)$, are solutions of the equation $z' = Az$. One can easily verify that $M_1 = \sup_{\|z\| \leq M, i \in \mathbb{Z}} \|W_i^1(z)\| < \infty$. Consider the following system:

$$v'(t) = Av(t) + f(v), t \neq \zeta_i(t_0),$$
$$\Delta v|_{t=\zeta_i(t_0)} = Bv(\zeta_i(t_0)) + W(v(\zeta_i(t_0))) + W_i^1(v(\zeta_i(t_0))), \qquad (10.3)$$

together with the system

$$z'(t) = Az(t) + f(z), t \neq \zeta_i(t_1),$$
$$\Delta|_{t=\zeta_i(t_1)} = Bz(\zeta_i(t_1)) + W(\zeta_i(t_1)), \tag{10.4}$$

where t_0, t_1 are the numbers under discussion.

Systems (10.3) and (10.4) are B-equivalent. That is, their solutions with the same initial condition coincide on the common domain if only $t \notin (\zeta_i(\widehat{t_0), \zeta_i}(t_1)], i \in \mathbb{Z}$. So, if $v(t), v(1) = z(1)$, is the solution of (10.3), then $v(t) = z(t)$ for all $t \notin (\zeta_i(\widehat{t_0), \zeta_i}(t_1)], i \in \mathbb{Z}$. For $v(t)$ we have that

$$v(t) = Z_2(t,1)v(1) + \int_{c1}^{t} Z_2(t,s)f(v(s))ds +$$
$$\sum_{1 \le \zeta_i < t} Z_2(t,\zeta_i(t_0))[W(v(\zeta_i(t_0)) + W_1(v(\zeta_i(t_0)))].$$

Thus,

$$\|\phi(t) - v(t)\| \le \|\phi(1) - v(1)\| \|Z_2(t,1)\| + \int_{c1}^{t} \|Z_2(t,s)\| L \|\phi(s) - v(s)\| ds +$$
$$\sum_{1 \le \zeta_j(t_0) < t} \|Z_2(t,\zeta_j(t_0))\| L \|\phi(\zeta_j(t_0)) - v(\zeta_j(t_0))\| +$$
$$\sum_{1 \le \zeta_j(t_0) < t} \|Z_2(t,\zeta_j(t_0))\| \|W_1(v(\zeta_i(t_0))\| \le$$
$$2MN + M_1 \frac{e^{\omega}}{1 - e^{-\omega}} + \int_{c1}^{t} N e^{-\omega(t-s)} L \|z(s) - v(s)\| ds +$$
$$\sum_{1 \le \zeta_j < t} N e^{-\omega(t-\zeta_j(t_0))} L \|v(\zeta_j(t_0)) - v(\zeta_j(t_0))\|.$$

Now, applying Lemma 2.5.1, we can find that

$$\|z(t) - v(t)\| \le (2MN + M_1 \frac{e^{\omega}}{1 - e^{-\omega}}) e^{(-\omega + NL + \ln(1+NL))(t-1)}.$$

The last inequality implies that $\|z(t) - v(t)\| < \epsilon$ if $t > \tilde{T}, t \notin [\zeta_i(\widehat{t_0), \zeta_i}(t_1)], i \ge 0$, where $\tilde{T} = 1 + \ln(\frac{\epsilon}{2MN + M_1 e^{\omega}(1-e^{-\omega})^{-1}})(-\omega + NL + \ln(1 + NL))^{-1}$, (we may assume that $\epsilon < 2M$). That is why, $z(t)(\epsilon, J)\phi(t)$ if $J = (t_1 + \tilde{T}, t_1 + \tilde{T} + E)$. The theorem is proved. □

Theorem 10.2.2. *Assume that conditions (C1)–(C6) are fulfilled. Then there exists a solution of (10.1), which is dense in $\mathcal{PC}$.*

Proof. Fix positive E, ϵ, and $t^* \in \Lambda$ such that the orbit of t^* is dense in Λ. Set $z_*(t) = z(t, t^*, z^*)$, $\|z^*\| < M$. Let us prove that $z_*(t)$ is the dense solution.

Consider an arbitrary solution $z(t) = z(t, t_0, z_0) \in \mathcal{PC}$. Consider an interval $J_1 = (0, E_1)$, where E_1 is an arbitrarily large positive number. By density of the orbit of t^* and uniform continuity of H, there exists a natural m such that

$$\|\zeta(t_1) - \zeta(t^*, m)\|_{J_1} < \epsilon, \tag{10.5}$$

where $\zeta(t^*, m) = \{\zeta_{i+m}(t^*)\}$.

We have

$$z_*(t+m) = Z_*(t+m, 1+m)z_*(1+m) + \int_{1+m}^{t+m} Z_*(t+m, u)f(z_*(u))du +$$

$$\sum_{1+m \le \zeta_i(t_0) < t+m} Z_*(t+m, \zeta_i(t_0))W(z_*(\zeta_i(t_0))) = Z_*(t+m, 1+m)z_*(1+m) +$$

$$\int_1^t Z_*(t, u)f(z_*(u+m))du + \sum_{1+m \le \zeta_i(t_0) < t+m} Z_*(t+m, \zeta_i(t_0))W(z_*(\zeta_i(t_0))),$$

and

$$z_1(t) = Z_1(t, 1)z_1(1) + \int_1^t Z_1(t, u)f(z_1(u))du + \sum_{1 \le \zeta_i(t_1) < t} Z_1(t, \zeta_i(t_1))W(z_1(\zeta_i(t_1))),$$

where Z_* and Z_1 are fundamental matrices corresponding to points t_* and t_1, respectively. Now, using the last two formulas, similarly to proof of Theorem 10.2.1, using (10.5) and the B-equivalence technique, we can find a sufficiently large number $E_1 > 2E$, and a natural number m such that $z_*(t+m)$ and $z_1(t)$ are ϵ-equivalent on $J = (E_1/2, E_1)$. The theorem is proved. □

Let $\overline{m} = \max_{|u| \le 1} \|e^{Au}\|$, $\underline{m} = \min_{|u| \le 1} \|e^{Au}\|$.

Condition $(C7)$ implies that $\eta = \min_{\|x\| \le M}(Bx + W(x)) > 0$.

From now on we make the assumption:

(C8) $L < \frac{\underline{m}\eta}{2\overline{m}M} \min\left(1, \frac{\underline{m}\overline{m}}{\overline{m}+\underline{m}}\right)$.

Theorem 10.2.3. *Assume that conditions (C1)–(C8) are fulfilled. Then (10.1) is sensitive on $\mathcal{PC}$.*

Proof. Fix a solution $z(t) = z(t, t_0, z_0)$, $t_0 \in \Lambda$, $z_0 \in \mathbb{R}^n$, and a positive δ. By sensitivity of H there exist $t_1 \in \Lambda, k > 0$, such that $|t_0 - t_1| < \delta$, $|\zeta_k(t_0) - \zeta_k(t_1)| \ge \delta_0$. Consequently, by uniform continuity of H, there exist numbers δ_1, δ_2, which do not depend on k and $t_0, t_1 \in \Lambda$, such that $|\zeta_{k-1}(t_0) - \zeta_{k-1}(t_1)| \ge \delta_1$,

$|\zeta_{k-2}(t_0) - \zeta_{k-2}(t_1)| \geq \delta_2$. Obviously, one can assume that $k > 3$. Moreover, uniform continuity of H implies that k can be an arbitrarily large number. Take arbitrary $z_1 \in \mathbb{R}^n$ and solution $z_1(t) = z(t, t_1, z_1)$.

Now, let us prove the sensitiveness through the solution $z_1(t)$.

Condition $(C8)$ implies that there exists a positive number ν such that $\frac{2\overline{m}M}{\underline{m}\eta} < \nu < \frac{\underline{m}\eta - 2\overline{m}ML}{\overline{m}}$.

We shall show that constants ϵ_0, ϵ_1 for Definition 10.2.1 can be taken equal to $\epsilon_0 = \min(\underline{m}\eta - \overline{m}(\nu + 2LM), \underline{m}\nu - \overline{m}2LM), \epsilon_1 = \min(\underline{\delta}, \frac{1}{2}(1 - \overline{\delta})$, where $\overline{\delta} = \max(\delta_0, \delta_1, \delta_2)), \underline{\delta} = \min(\delta_0, \delta_1, \delta_2)$. One can easily see that among numbers k and $k-1$ there exists one , let us say k itself, such that $|\zeta_k(t_0) - \zeta_k(t_1)| \geq \epsilon_1$ and interval $[\zeta_k(t_0) - \epsilon_1, \zeta_k(t_0))$ does not have points of discontinuity from $\zeta(t_0)$ and $\zeta(t_1)$.

Assume that $\|z(\zeta_k(t_0)) - z_1(\zeta_k(t_0))\| < \nu$. Then, for $t \in [\zeta_k(t_0), \zeta_k(t_1)]$,

$$z(t) = \mathrm{e}^{A(t-\zeta_k(t_0))}(\mathcal{I} + B)z((\zeta_k(t_0)) + \int_{\zeta_k(t_0)}^{t} \mathrm{e}^{A(t-s)} f(z(s))ds +$$

$$\mathrm{e}^{A(t-\zeta_k(t_0))} W(z((\zeta_k(t_0))),$$

$$z_1(t) = \mathrm{e}^{A(t-\zeta_k(t_0))} z_1((\zeta_k(t_0)) + \int_{\zeta_k(t_0)}^{t} \mathrm{e}^{A(t-s)} f(z_1(s))ds.$$

We have that

$$\|z(t) - z_1(t)\| = \|\mathrm{e}^{A(t-\zeta_k(t_0))}[Bz(\zeta_k(t_0)) + W(z(\zeta_k(t_0)))] + \mathrm{e}^{A(t-\zeta_k(t_0))}[z((\zeta_k(t_0)) -$$

$$z_1((\zeta_k(t_0))] + \int_{\zeta_k(t_0)}^{t} \mathrm{e}^{A(t-s)}(f(z(s)) - f(z_1(s)))ds\| \geq \underline{m}\eta - \overline{m}(\nu + 2LM) \geq \epsilon_0.$$

If $\|z(\zeta_k(t_0)) - z_1(\zeta_k(t_0))\| > \nu$, then, for $t \in [\zeta_k(t_0) - \epsilon_1, \zeta_k(t_0))$,

$$z(t) = \mathrm{e}^{A(t-\zeta_k(t_0))} z((\zeta_k(t_0)) + \int_{\zeta_k(t_0)}^{t} \mathrm{e}^{A(t-s)} f(z(s))ds,$$

$$z_1(t) = \mathrm{e}^{A(t-\zeta_k(t_0))} z_1((\zeta_k(t_0)) + \int_{\zeta_k(t_0)}^{t} \mathrm{e}^{A(t-s)} f(z_1(s))ds.$$

and $\|z(t) - z_1(t)\| \geq \underline{m}\nu - \overline{m}2LM \geq \epsilon_0$. The theorem is proved. □

On the basis of Theorems 10.2.1–10.2.3, we can conclude that (10.1) admits the Devaney's chaos.

It seems natural to consider the chaos only for uniformly bounded solutions on $[0, \infty)$, since the domain of chaos is always assumed to be a compact set, but we consider chaotic properties of all solutions, since the chaotic scenario for these unbounded solutions starts at the moment they reach the region where solutions from $\mathcal{PCA}$ are placed. This set is a chaotic attractor as it is easily seen that $\mathcal{PCA}$ admits defined above all ingredients of Devaney's chaos.

10.3 Shadowing Property

In this part of the chapter, we give definitions of shadowing property for the flow of system (10.1) and prove it for this system if the generator map has the property. A corollary of the result for a map H with the hyperbolic set Λ is obtained.

Assume that the generator map, $H(t)$, is defined in a neighborhood of the unit interval I.

The following definitions are from [55, 122, 131, 134] and are adapted for our system.

A sequence $\{\kappa_i\}_0^N, N \leq \infty$, is said to be a *true* trajectory of H, if $\kappa_0 \in \Lambda$ and $\kappa_{i+1} = H(\kappa_i), 0 \leq i < N$.

A sequence $\{\pi_i\}_0^N, N \leq \infty$, is said to be a *$\kappa$-pseudo-orbit*, $\kappa > 0$, of H, if $|\pi_{i+1} - H(\pi_i)| < \kappa$, and $|p_i - \lambda| < \kappa$ for all $0 \leq i < N$, and $\lambda \in \Lambda$.

The true orbit $\{\kappa_i\}_0^N$ δ-shadows the pseudo-orbit $\{\pi_i\}_0^N$ if $|\kappa_i - \pi_i| < \delta$ for all i.

A sequence $\{z_i\}_0^N$ is said to be a *true discrete orbit* of (10.1) if $z_{i+1} = z(\zeta_{i+1}, \zeta_i, z_i)$, where $\zeta_i = i + \kappa_i$ for all $0 \leq i < N$. Let δ be a positive number, and k a positive integer. A sequence y_{ik} such that $0 \leq ik \leq N$ if $N < \infty$, and $i \geq 0$, if $N = \infty$, is said to be a *discrete δ-pseudo-orbit* for the problem (10.1) with associated sequence $\{p_i\}_0^N$ if $\|y_{(i+1)k} - w(p_{(i+1)k})\| < \delta$ for all admissible i, and the solution $w(t)$ of the initial value problem

$$
\begin{aligned}
w'(t) &= Aw(t) + f(w),\\
\Delta|_{t=p_i} w &= Bw(p_i) + W(w(p_i)),\\
w(p_{ik}) &= y_{ik}.
\end{aligned}
\tag{10.6}
$$

A discrete δ-pseudo-orbit y_{ik} of problem (10.1) is said to be ϵ-shadowed by a true orbit $\{z_i\}_0^N$ of (10.1) if $\|z_{ik} - y_{ik}\| < \epsilon$, and $|\zeta_{ik} - p_{ik}| < \epsilon$ for all i such that $0 \leq ik \leq N$ if $N < \infty$, and $i \geq 0$, if $N = \infty$. Consider the logistic function $h(x, \mu) \equiv \mu x(1 - x)$ with coefficient $\mu = 3.8$. It is proved in [76] that for $\epsilon = 10^{-8}, N = 10^7, p_0 = 0.4$, the pseudo-orbit $p_i, i = 0$ to N, is ϵ-shadowed by a true orbit, if $\delta = 3 \times 10^{-14}$. Several values of μ were claimed to be proper for the shadowing. Taking into account this result as well as results from [40, 55, 61, 120, 134] the following assertion is very useful.

Theorem 10.3.1. *Assume that conditions (C1)–(C6) are fulfilled. Then, given $\epsilon > 0$, there exists $0 < \delta < \epsilon$ and a positive integer k such that a δ-pseudo-orbit y_{ik} of problem (10.1) is ϵ-shadowed by a true orbit $\{z_i\}_0^N$ of (10.1) if $p_i = i + \pi_i$, and π_i is δ-shadowed by $\{\kappa_i\}_0^N$.*

Proof. Fix positive ϵ and nonnegative integer i. We assume that $\|z_{ik} - y_{ik}\| < \epsilon$, and we will find δ and k, such that $\|z_{(i+1)k} - y_{(i+1)k}\| < \epsilon$. Assume, without loss of generality, that $\zeta_{ik} < p_{ik}$, and let $z(t) = z(t, \zeta_{ik}, z_{ik})$. We have that

$$\|z(p_{ik}) - y_{ik}\| \le \|z(p_{ik}) - z_{ik}\| + \|z_{ik} - y_{ik}\| = \|e^{A(p_{ik}-\zeta_{ik})}z_{ik}$$
$$+ \int_{\zeta_{ik}}^{p_{ik}} e^{A(t-s)} f(z(s))ds\| + \|z_{ik} - y_{ik}\| \le$$
$$\|[\mathcal{I} - e^{A(p_{ik}-\zeta_{ik})}]\|\|z_{ik}\| + \delta N M_0 + \epsilon = \delta\phi(\delta) + \epsilon,$$

where $\phi(s)$ is a bounded function.

Similarly to the proof of Theorem 10.2.1, we find that (10.1) is B-equivalent to the following system:

$$\begin{aligned} v'(t) &= Av(t) + f(v), \\ \Delta v|_{t=p_i} &= Bv(p_i) + W(v(p_i)) + \tilde{W}_i^1(v(p_i)), \\ v(t_0) &= z_0,\ (t_0, z_0) \in \Lambda \times \mathbb{R}^n, \end{aligned} \tag{10.7}$$

with $M_2 = \sup_{\|z\| \le M, i \in \mathbb{Z}} \|\tilde{W}_i^1(z)\| < \infty$.

Then we can obtain that

$$\|z(t) - w(t)\| \le [N(\delta\phi(\delta) + \epsilon) + M_2 \frac{e^{\omega}}{1 - e^{-\omega}}]e^{(-\omega + NL + \ln(1+NL))(t-1)},$$

if $t \notin \widehat{[p_i, \zeta_i)}$. Now, choose k sufficiently large, and δ small for the right-hand side of the last inequality to be less than $\frac{\epsilon}{3}$ at $t = (i+1)k - 1$, and $\delta \max(1, \phi(\delta)) < \frac{\epsilon}{3}$. Then $\|z_{(i+1)k} - y_{(i+1)k}\| < \|z_{(i+1)k} - z(p_{(i+1)k})\| + \|z(p_{(i+1)k}) - w(p_{(i+1)k})\| + \|y_{(i+1)k} - w(p_{(i+1)k})\| < \epsilon$. The theorem is proved. □

Now, by using the Shadowing Theorem [55, 122, 131] one can easily prove that the following assertion is true.

Theorem 10.3.2. *Assume that conditions (C1)–(C6) are fulfilled and H has a compact positively invariant hyperbolic set $\Lambda \subset I$. Then, given $\epsilon > 0$, there exist $0 < \delta < \epsilon$, and a positive integer k such that a δ-pseudo-orbit $\{y_{ik}\}_0^\infty$, of problem (10.1) is ϵ-shadowed by a true orbit $\{z_i\}_0^\infty$ of (10.1) if $\pi_i = p_i - i, i \ge 0$, is a δ-pseudo-orbit of H.*

10.4 Simulations

Consider the following initial value problem

$$\begin{aligned} x_1' &= 2/5x_2 + l\sin^2 x_2, \\ x_2' &= 2/5x_1 + l\sin^2 x_1, t \ne \zeta_i(t_0), \\ \Delta x_1|_{t=\zeta_i(t_0)} &= -\frac{4}{3}x_1, \\ \Delta x_2|_{t=\zeta_i(t_0)} &= -\frac{4}{3}x_2 + W(x_2), \end{aligned} \tag{10.8}$$

where $W(s) = 1 + s^2$, if $|s| \leq l, l$ is a positive constant, and $W(s) = 1 + l^2$, if $|s| > l$. One can easily see that all the functions are lipschitzian with a constant proportional to l. The matrices of coefficients

$$A = \begin{pmatrix} 0 & 2/5 \\ 2/5 & 0 \end{pmatrix}, \quad B = \begin{pmatrix} -4/3 & 0 \\ 0 & -4/3 \end{pmatrix}$$

commute, and the eigenvalues of the matrix

$$A + \text{Ln}(\mathcal{I} + B) = \begin{pmatrix} -\ln 3 & 2/5 \\ 2/5 & -\ln 3 \end{pmatrix}$$

are negative: $\lambda_{1,2} = -\ln 3 \pm 2/5 < 0$.

The results of the last section make possible the following appropriate simulations.

Choose $\mu = 3.8$ and $l = 10^{-2}$ in (10.8) and consider the solution $x(t) = (x_1, x_2)$ with initial moments $t_0 = 7/9$ and the initial value $x(t_0) = (0.005, 0.002)$.

If one consider the sequence $(x_1(n), x_2(n)), n = 1, 2, 3, \ldots, 75000$, in x_1, x_2-plane, then the attractor can be seen, Fig. 10.1. To approve that the attractor is chaotic, we verify the conditions of the chaotic theorems in the following way. If $|s| \leq l$, then $-\frac{4}{3}s + W(s) = s^2 - \frac{4}{3}s + 1$, and it is never equal to zero. If $|s| > l$, then $-\frac{4}{3}s + W(s) = l^2 - \frac{4}{3}s + 1$. For the last expression to be zero, we need, $s = \frac{3}{4}(1 + l^2)$. From the Figure, it is seen that the second coordinate takes values between 0.32 and 0.42. This is the region where $-\frac{4}{3}s + W(s)$ does not have zeros. All the other conditions required by theorems of this chapter could be easily checked with sufficiently small coefficient l.

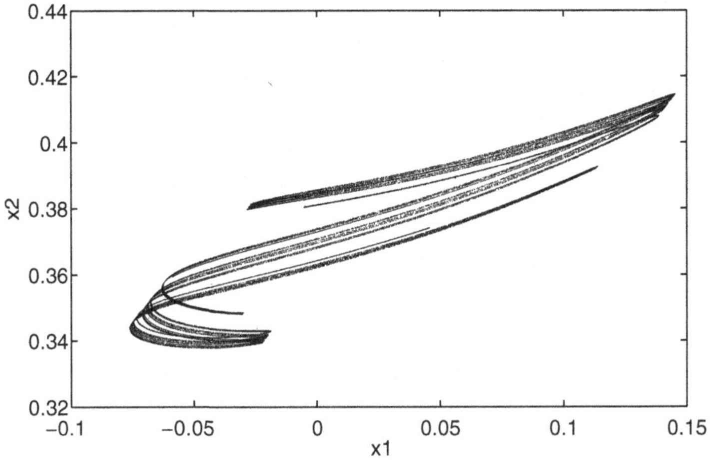

Fig. 10.1 The chaotic attractor by a stroboscopic sequence $(x_1(n), x_2(n))$, $1 \leq n \leq 75{,}000$

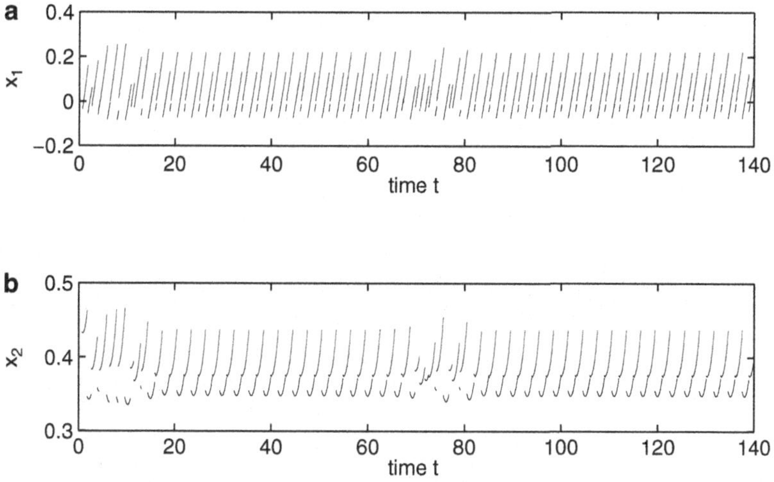

Fig. 10.2 The intermittency of the both coordinates $x_1(t), x_2(t)$ is observable

Now, consider (10.8) with $\mu = 3.8282$. Then the phenomenon of intermittency, i.e., irregular switching between periodic and chaotic behavior, for the solution $x(t)$ can be observed in Fig.10.2. The coefficient's value is such that the logistic map admits intermittency [62].

Notes

The investigation of the last chapter is inspired by the discontinuous dynamics of the neural information processing in the brain, information communication, and population dynamics [70,88,99,100,103,108,160]. While there are many interesting papers concerned with the complex behavior generated by impulses, the rigorous theory of chaotic impulsive systems remains far from being complete. Our goal is to develop further the theoretical foundations of this area of research. The complex dynamics is obtained using Devaney's definition for guidance. The main results of this chapter are published in [11]. There simulations for a pendulum are given. Applications of the present approach to the analysis of the cardiovascular system were considered in [12, 16]. More of our results on chaos excitability can be found in [8–10].

References

1. E. Akalin, M.U. Akhmet, The principles of B-smooth discontinuous flows, Comput. Math. Appl., 49 (2005) 981–995.
2. M.U. Akhmet, On the general problem of stability for impulsive differential equations, J. Math. Anal. Appl., 288 (2003) 182–196.
3. M.U. Akhmet, On the smoothness of solutions of impulsive autonomous systems, Nonlinear Anal.: TMA, 60 (2005a) 311–324.
4. M.U. Akhmet, Perturbations and Hopf bifurcation of the planar discontinuous dynamical system, Nonlinear Anal.: TMA, 60 (2005b) 163–178.
5. M.U. Akhmet, Integral manifolds of differential equations with piecewise constant argument of generalized type, Nonlinear Anal.: TMA, 66 (2007a) 367–383.
6. M.U. Akhmet, On the reduction principle for differential equations with piecewise constant argument of generalized type, J. Math. Anal. Appl., 336 (2007b) 646–663.
7. M.U. Akhmet, Almost periodic solutions of differential equations with piecewise constant argument of generalized type, Nonlinear Anal.: HS, 2 (2008) 456-467.
8. M.U. Akhmet, Devaney's chaos of a relay system, Commun. Nonlinear Sci. Numer. Simul., 14 (2009a) 1486–1493.
9. M.U. Akhmet, Li-Yorke chaos in the impact system, J. Math. Anal. Appl., 351 (2009b), 804–810.
10. M.U. Akhmet, Dynamical synthesis of quasi-minimal sets, Int. J. Bifurcation Chaos, 19, no. 7 (2009c) 1–5.
11. M.U. Akhmet, Shadowing and dynamical synthesis, Int. J. Bifurcation Chaos, Int. J. Bifurcation Chaos, 19, no. 10 (2009d) 1–8.
12. M.U. Akhmet, The complex dynamics of the cardiovascular system. Nonlinear Anal.: TMA, 71 (2009e) e1922–e1931.
13. M.U. Akhmet, D. Arugaslan, Bifurcation of a non-smooth planar limit cycle from a vertex, Nonlinear Anal.: TMA, 71 (2009) e2723–e2733.
14. M.U. Akhmet, D. Arugaslan, M. Beklioglu, Impulsive control of the population dynamics, Proceedings of the Conference on Differential and Difference Equations at the Florida Institute of Technology, August 1–5, 2005, Melbourne, Florida, Editors: R.P. Agarval and K. Perera, Hindawi Publishing Corporation, 2006, 21–30.
15. M.U. Akhmet, D. Arugaslan, M. Turan, Hopf bifurcation for a 3D Filippov system, Dyn. Contin. Discrete Impuls. Syst., Ser. A, 16 (2009) 759–775.
16. M.U. Akhmet, G.A. Bekmukhambetova, A prototype compartmental model of the blood pressure distribution, Nonlinear Anal.: RWA, 11 (2010), 1249–1257.
17. M.U. Akhmet, C. Buyukadali, Differential equations with a state-dependent piecewise constant argument, Nonlinear Analysis: TMA, 72 (2010), 4200–4211.
18. M.U. Akhmet, M. Kirane, M.A. Tleubergenova, G.W. Weber, Control and optimal response problems for quasi-linear impulsive integro-differential equations, Eur. J. Operational Res., 169 (2006) 1128–1147.

19. M.U. Akhmet, M. Turan, The differential equations on time scales through impulsive differential equations, Nonlinear Anal.: TMA, 65 (2006) 2043–2060.
20. M.U. Akhmet, M. Turan, Differential equations on variable time scales, Nonlinear Anal.: TMA, 70 (2009a) 1175–1192.
21. M.U. Akhmet, M. Turan, Bifurcation of 3D discontinuous cycles, Nonlinear Anal.: TMA, 71 (2009b) e2090–e2102.
22. M.U. Akhmetov, On periodic solutions of certain systems of differential equations, Vestn. Kiev. Univer., Mat. i Mekh. (russian), 24 (1982) 3–7.
23. M.U. Akhmetov, Periodic solutions of non-autonomous systems of differential equations with impulse action in the critical case (russian), Izv. Akad. Nauk Kazakh. SSR, Seria Fiz.-Mat., (1991a) no. 3, 62–65.
24. M.U. Akhmetov, On the existence of higher-order B-derivatives of solutions of impulse systems with respect to initial data (russian), Izv. Akad. Nauk Kazakh. SSR, Seria Fiz.-Mat., (1991b) no. 1, 15–17.
25. M.U. Akhmetov, Asymptotic representation of solutions of regularly perturbed systems of differential equations with a non classical right-hand side. Ukrainian Math. J., 43 (1991c) 1298–1304.
26. M.U. Akhmetov, Periodic solutions of systems of differential equations with a non classical right-hand side containing a small parameter (russian), TIC: Collection: asymptotic solutions of non linear equations with small parameter. 1991d, 11–15. UBL: Akad. Nauk Ukr. SSR, Inst. Mat., Kiev.
27. M.U. Akhmetov, On the smoothness of solutions of differential equations with a discontinuous right-hand side, Ukrainian Math. J., 45 (1993) 1785–1792.
28. M.U. Akhmetov, On the method of successive approximations for systems of differential equations with impulse action at nonfixed moments of time (russian), Izv. Minist. Nauki Vyssh. Obraz. Resp. Kaz. Nats. Akad. Nauk Resp. Kaz. Ser. Fiz.-Mat. (1999) no. 1, 11–18.
29. M.U. Akhmetov, R.F. Nagaev, Periodic solutions of a nonlinear impulse system in a neighborhood of a generating family of quasiperiodic solutions, Differ. Equ., 36 (2000) 799–806.
30. M.U. Akhmetov, N.A. Perestyuk, On the almost periodic solutions of a class of systems with impulse effect (russian), Ukr. Mat. Zh., 36 (1984) 486–490.
31. M.U. Akhmetov, N.A. Perestyuk, Almost periodic solutions of sampled-data systems. Ukr. Mat. Zh. (russian), 39 (1987) 74–80.
32. M.U. Akhmetov, N.A. Perestyuk, On motion with impulse actions on a surfaces (russian), Izv.-Acad. Nauk Kaz. SSR, Ser. Fiz.-Mat., 1 (1988) 111–114.
33. M.U. Akhmetov, N.A. Perestyuk, The comparison method for differential equations with impulse action, Differ. Equ., 26 (1990) 1079–1086.
34. M.U. Akhmetov, N.A. Perestyuk, Asymptotic representation of solutions of regularly perturbed systems of differential equations with a non-classical right-hand side, Ukrainian Math. J., 43 (1991) 1209–1214.
35. M.U. Akhmetov, N.A. Perestyuk, Differential properties of solutions and integral surfaces of nonlinear impulse systems, Differ. Equ., 28 (1992a) 445–453.
36. M.U. Akhmetov, N.A. Perestyuk, Periodic and almost periodic solutions of strongly nonlinear impulse systems, J. Appl. Math. Mech., 56 (1992b) 829–837.
37. M.U. Akhmetov, N.A. Perestyuk, On a comparison method for pulse systems in the space $\mathbb{R}^n$, Ukr. Math. J. 45 (1993) 826–836.
38. A.A. Andronov, A.A. Vitt, C.E. Khaikin, Theory of oscillations, Pergamon, Oxford, 1966.
39. D.V. Anosov, V.I. Arnold, Dynamical Systems, Springer, Berlin, 1994.
40. D.V. Anosov, Geodesic flows and closed Riemannian manifolds with negative curvature, Proc. Steklov Inst. Math. 90 (1967).
41. J. Awrejcewicz, C.H. Lamarque, Bifurcation and chaos in nonsmooth mechanical systems, World Scientific, Singapore, 2003.
42. G. Aymerich, Sulle oscillazioni autosostenute impulsivamente, Rend. Semin. Fac. Sci. Univ. Cagliari, 22 (1952) 34–37.
43. V.I. Babitsky, Theory of vibro-impact systems and applications, Springer, Berlin, 1998.

44. A. Balanov, N. Janson, D. Postnov, O. Sosnovtseva, Synchronization: from simple to complex, Springer, Berlin, 2009.
45. D.D. Bainov, P.S. Simeonov, Stability under persistent disturbances for systems with impulse effect, J. Math. Anal. Appl., 109 (1985) 546–563.
46. D.D. Bainov, P.S. Simeonov, Impulsive differential equations: asymptotic properties of solutions, World Scientific, Singapore, New Jersey, London, 1995.
47. D.D. Bainov, V. Govachev, Impulsive differential equations with a small parameter, World Scientific, Singapore, New Jersey, London, Hong Kong, 1994.
48. N.N. Bautin, E.A. Leontovich, Methods and rules for the qualitative study of dynamical systems on the plane (russian), Nauka, Moscow, 1990.
49. M. Benedicks, L. Carleson, The dynamics of the Henon map, Ann. Math., 133 (1991) 73–169.
50. M. di Bernardo, C.J. Budd, A.R. Champneys, P. Kowalczyk, Piecewise-smooth dynamical systems, Springer, London, 2008a.
51. M. di Bernardo, A. Nordmark, G. Olivar, Discontinuity-induced bifurcations of equilibria in piecewise-smooth and impacting dynamical systems, Phys. D, 237 (2008b) 119–136.
52. R. Bellman, Mathematical methods in medicine, World Scientific, Singapore, 1983.
53. I.I. Blekhman, Synchronization of dynamical systems (russian), Nauka, Moscow, 1971.
54. E.M. Bonotto, M. Federson, Limit sets and the Poincaré-Bendixson theorem in impulsive semidynamical systems, J. Diff. Eqs., 244 (2008) 2334–2349.
55. R. Bowen, ω-limit sets for Axiom A diffeomorfisms, J. Diff. Eqs., 18 (1975) 333–339.
56. B. Brogliato, Nonsmooth impact mechanics, Springer, London, 1996.
57. B. Brogliato, Impacts in mechanical systems – Analysis and modeling, Springer, New York, 2000.
58. D. Chillingwirth, Differential topology with a view to applications, Pitman, London, 1978.
59. E.A. Coddington, N. Levinson, Theory of Ordinary Differential Equations, McGraw-Hill, New York, 1955.
60. C. Corduneanu, Principles of differential and integral equations, Chelsea Publishing Co., Bronx, NJ, 1977.
61. E.M. Coven, I. Kan, J.A. Yorke, Pseudo-Orbit Shadowing in the Family of Tent Maps, Trans. Am. Math. Soc., 308 (1988) 227–241.
62. R. Devaney, An introduction to chaotic dynamical systems, Addison-Wesley, Menlo Park, CA, 1990.
63. A.B. Dishliev, D.D. Bainov, Sufficient conditions for absence of 'beating' in systems of differential equations with impulses, Appl. Anal., 18 (1984) 67–73.
64. A.B. Dishliev, D.D. Bainov, Differentiability on a parameter and initial condition of the solution of a system of differential equations with impulses. sterreich. Akad. Wiss. Math.-Natur. Kl. Sitzungsber. II 196 (1987) 69–96.
65. A. Domoshnitsky, M. Drakhlin, E. Litsyn, Nonoscillation and positivity of solutions to first order state-dependent differential equations with impulses in variable moments, J. Diff. Eqs., 228 (2006) 39–48.
66. M. Feckan, Bifurcation of periodic and chaotic solutions in discontinuous systems. Arch. Math. (Brno), 34 (1998) 73–82.
67. M.I. Feigin, Doubling of the oscillation period with C-bifurcations in piecewise continuous systems (Russian), J. Appl. Math. Mech., 38 (1974) 810–818.
68. A.F. Filippov, Differential equations with discontinuous right-hand sides, Kluwer, Dordrecht, 1988.
69. M. Frigon, D. O'Regan, Impulsive differential equations with variable times, Nonlinear Anal.: TMA, 26 (1996) 1913–1922.
70. L. Glass, M.C. Mackey, A simple model for phase locking of biological oscillators, J. Math. Biol., 7 (1979) 339–352.
71. J. Guckenheimer, P.J. Holmes, Nonlinear oscillations, dynamical systems and bifurcations of vector fields, Springer, New-York, 1083.
72. J. Guckenheimer, R.F. Williams, Structural stability of Lorentz attractors, Publ. Math. IHES, 50 (1979) 59–72.

73. V. Gullemin, A. Pollack, Differential topology, Prentice-Hall, New Jersey, 1974.
74. A.M. Gupal, V.I. Popadinets, Notes on formulas of differentiability of 'discontinuous solutions' of systems of ordinary differential equations on initial values and parameters (russian), Kibernetika, 4 (1974) 148–149.
75. A. Halanay, D. Wexler, Qualitative theory of impulsive systems (romanian), Edit. Acad. RPR, Bucuresti, 1968.
76. S.M. Hammel, J.A. Jorke, C. Grebogi, Do numerical orbits of chaotic dynamical processes represent true orbits? J. Complexity, 3 (1987) 136–145.
77. P. Hartman, Ordinary Differential Equations, Wiley, New York, 1964.
78. C.S. Hcu, W.H. Cheng, Applications of the theory of impulsive parametric excitation and new treatment of general parametric excitations problems, Trans. ASME, 40 (1973) 2174–2181.
79. P.J. Holmes, The dynamics of repeated impacts with a sinusoidal vibrating table, J. Sound Vib., 84 (1982) 173–189.
80. E. Hopf, Abzweigung einer periodishen Losung von einer stationaren Losung eines Differential systems, Ber. Math.-Phys. Sachsische Academie der Wissenschaften, Leipzig, 94 (1942) 1–22.
81. F.C. Hoppensteadt, C.S. Peskin, Mathematics in Medicine and in the Life Sciences, Springer, New York, 1992.
82. S.C. Hu, V. Lakshmikantham, S. Leela, Impulsive differential systems and the pulse phenomena, J. Math. Anal. Appl., 137 (1989) 605–612.
83. G. Iooss, D.D. Joseph, Bifurcation of maps and applications, Springer, New York, 1980.
84. M.V. Jakobson, Absolutely continuous invariant measures for one parameter families of one-dimensional maps, Commun. Math. Phys., 81 (1981) 39–88.
85. Qi Jiangang, Fu Xilin, Existence of limit cycles of impulsive differential equations with impulses at variable times. Nonlinear Anal.: TMA, 44 (2001) 345–353.
86. A. Katok, J.-M. Strelcyn, F. Ledrappier, F. Przytycki, Invariant Manifolds, Entropy and Billiards; Smooth Maps with Singularities, Lecture Notes in Mathematics, 1222, Springer, Berlin, 1986.
87. S. Kaul, On impulsive semidynamical systems, J. Math. Anal. Appl., 150 (1990) 120–128.
88. A. Khadra, X. Liu, X. Shen, Impulsive control and synchronization of spatiotemporal chaos, Chaos, Solitons and Fractals, 26 (2005) 615–636.
89. A.E. Kobrinskii, A.A. Kobrinskii, Vibro-shock systems (russian), Nauka, Moscow, 1971.
90. A.N. Kolmogorov, On the Skorokhod convergence (russian), Teor. Veroyatn. i Prim., 1 (1956) 239–247.
91. N.M. Krylov, N.N. Bogolyubov, Introduction to nonlinear mechanics, Acad. Nauk Ukrainy, Kiev, 1937.
92. M. Kunze, Non-Smooth Dynamical Systems, Lecture Notes in Mathematics, Vol. 1744, Springer, Berlin, 2000.
93. M. Kunze, T. Küpper, Qualitative bifurcation analysis of a non-smooth friction-oscillator model, Z. Angew. Math. Phys., 48 (1997) 87–101.
94. Y. Kuramoto, Chemical oscillations, Springer, Berlin, 1984.
95. V. Lakshmikantham, D.D. Bainov, P.S. Simeonov, Theory of impulsive differential equations, World Scientific, Singapore, NJ, London, Hong Kong, 1989.
96. V. Lakshmikantham, S. Leela, S. Kaul, Comparison principle for impulsive differential equations with variable times and Stability theory, Nonlinear Anal.: TMA, 22 (1994) 499–503.
97. V. Lakshmikantham, X. Liu, On quasistability for impulsive differential equations, Nonlinear Anal.: TMA, 13 (1989) 819–828.
98. S. Lefschetz, Differential equations: Geometric theory, Interscience Publishers, New York, 1957.
99. W. Lin, R. Jiong, Chaotic dynamics of an integrate-and-fire circuit with periodic pulse-train input., IEEE Trans. Circuits Syst. I Fund. Theory Appl., 50 (2003) 686–693.
100. W. Lin, Description of complex dynamics in a class if impulsive differential equations, Chaos, Solutions and Fractals, 25 (2005) 1007–1017.
101. L. Liu, J. Sun, Existence of periodic solution for a harvested system with impulses at variable times, Phys. Lett. A, 360 (2006) 105–108.

102. X. Liu, R. Pirapakaran, Global stability results for impulsive differential equations, Appl. Anal., 33 (1989) 87–102.
103. A.C.J. Luo, Global transversality, resonance and chaotic dynamics, World Scientific, Hackensack, NJ, 2008.
104. A.M. Lyapunov, Probléme général de la stabilité du mouvement, Princeton University Press, Princeton, N.J., 1949.
105. I.G. Malkin, Theory of stability of motion, U.S. Atomic Energy Commission, Office of Technical Information, 1958.
106. J.E. Marsden, M. McCracken, The Hopf bifurcation and its applications, Appl. Math. Sci., Vol. 19, Springer, New York, 1976.
107. N. Minorsky, Nonlinear Oscillations, D. Van Nostrand Company, Inc. Princeton, London, New York, 1962.
108. R.E. Mirollo, S.H. Strogatz, Synchronization of pulse-coupled biological oscillators, SIAM J. Appl. Math., 50 (1990) 1645–1662.
109. E. Mosekilde, Zh. Zhusubalyev, Bifurcations and chaos in piecewise-smooth dynamical systems, World Scientific, River Edge, NJ, 2003.
110. A.D. Myshkis, On asymptotic stability of the rough stationary points of the discontinuous dynamic systems on plane, Autom. Remote Contrl., 62 (2001) 1428–1432.
111. A.D. Myshkis, A.M. Samoilenko, Systems with impulses at fixed moments of time (russian), Math. Sb., 74 (1967) 202–208.
112. R.F. Nagaev, Periodic solutions of piecewise continuous systems with a small parameter (russian), Prikl. Mat. Mech., 36 (1972) 1059–1069.
113. R.F. Nagaev, Mechanical processes with repeated and decaying impacts (russian), Nauka, Moscow, 1985.
114. R.F. Nagaev, Dynamics of synchronising systems. Springer, Berlin, 2003.
115. R.F. Nagaev, D.G. Rubisov, Impulse motions in a one-dimensional system in a gravitational force field, Soviet Appl. Mech., 26 (1990) 885–890.
116. Yu.I. Neimark, The method of point transformations in the theory of nonlinear oscillations (russian), Nauka, Moscow, 1972.
117. V.V. Nemytskii, V.V. Stepanov, Qualitative theory of Differential Equations, Princeton University Press, Princeton, New Jersey, 1966.
118. A.B. Nordmark, Existence of periodic orbits in grazing bifurcations of impacting mechanical oscillators, Nonlinearity, 14 (2001) 1517–1542.
119. H.E. Nusse, E. Ott, J.A. Yorke, Border-collision bifurcations: an explanation for observed bifurcation phenomena, Phys. Rev. E, 49 (1994) 1073–1076.
120. H.E. Nusse, J.A. Yorke, Is Every Approximate Trajectory of Some Process Near an Exact Trajectory of a Nearby Process? Commun. Math. Phys., 114 (1988) 363–379.
121. M. Oestreich, N. Hinrichs, K. Popp, C.J. Budd, Analytical and experimental investigation of an impact oscillator. Proceedings of the ASME 16th Biennal Conf. on Mech. Vibr. and Noise, DETC97VIB-3907: 1–11, 1997.
122. K. Palmer, Shadowing in dynamical systems: Theory and Applications, Kluwer, Dordrecht, 2000.
123. T. Pavlidis, A new model for simple neural nets and its application in the design of a neural oscillator, Bull. Math. Biophys., 27 (1965) 215–229.
124. T. Pavlidis, Stability of a class of discontinuous dynamical systems, Inform. Contrl., 9 (1966) 298–322.
125. F. Peterka, Part I: Theoretical analysis of n-multiple $(1/n)$-impact solutions, CSAV Acta Technica, 26 (1974) 462–473.
126. F. Pfeiffer, Multibody systems with unilateral constraints (russian), J. Appl. Math. Mech., 65 (2001) 665–670.
127. F. Pfeiffer, Chr. Glocker, Multibody dynamics with unilateral contacts. Wiley, New York, 1996.
128. N.A. Perestyuk, V.N. Shovkoplyas, Periodic solutions of nonlinear impulsive differential equations, Ukr. Mat. Zh., 31 (1979) 517–524.

129. A. Pikovsky, Synchronization : a universal concept in nonlinear sciences, Cambridge University Press, Cambridge, 2001.
130. V.N. Pilipchuk, R.A. Ibrahim, Dynamics of a two-pendulum model with impact interaction and an elastic support, Nonlinear Dynam., 21 (2000) 221–247.
131. S.Yu. Pilugin, Shadowing in dynamical systems, Springer, Berlin, 1999.
132. H. Poincaré, Les méthodes nouvelles de la mécanique céleste, 2,3, Gauthier-Villars, Paris, 1892.
133. L.D. Pustylnikov, Stable and oscillating motions in non-autonomous dynamical systems, Trans. Math. Soc., 14 (1978) 1–10.
134. C. Robinson, Dynamical Systems: stability, symbolic dynamics, and chaos, CRC, Boca Raton, Ann Arbor, London, Tokyo, 1995.
135. V.F. Rozhko, Lyapunov stability in discontinuous dynamic systems, (russian), Diff. Eqs., 11 (1975) 761–766.
136. V.F. Rozhko, On a class of almost periodic motions in systems with shocks (russian), Diff. Eqs., 11 (1972) 2012–2022.
137. A.M. Samoilenko, N.A. Perestyuk, Second N.N. Bogolyubov's theorem for impulsive differential equations (russian), Differentsial'nye uravneniya, 10 (1974) 2001–2010.
138. A.M. Samoilenko, N.A. Perestyuk, Stability of solutions of impulsive differential equations (russian), Differentsial'nye uravneniya, 13 (1977) 1981–1992.
139. A.M. Samoilenko, N.A. Perestyuk, Periodic solutions of weakly nonlinear impulsive differential equations (russian), Differentsial'nye uravneniya, 14 (1978) 1034–1045.
140. A.M. Samoilenko, N.A. Perestyuk, On stability of solutions of impulsive systems (russian), Differentsial'nye uravneniya, 17 (1981) 1995–2002.
141. A.M. Samoilenko, N.A. Perestyuk, Differential Equations with impulsive actions (russian), Vishcha Shkola, Kiev, 1987.
142. A.M. Samoilenko, N.A. Perestyuk, Impulsive Differential Equations, World Scientific, Singapore, 1995.
143. G. Sansone, Sopra una equazione che si presenta nelle determinazioni della orbite in un sincrotrone, Rend. Accad. Naz. Lincei, 8 (1957) 1–74.
144. S.W. Shaw, P.J. Holmes, Periodically forced linear oscillator with impacts: Chaos and long-period motions, Phys. Rev. Lett., 51 (1983a) 623–626.
145. S.W. Shaw, P.J. Holmes, A periodically forced piecewise linear oscillator, J. Sound Vibr., 90 (1983b) 129–155.
146. Ya. G. Sinai, What is ... a billiard? Notices Am. Math. Soc., 51 (2004) 412–413.
147. A.V. Skorokhod, Limit theorems for random processes, (russian), Teor. Veroyatnost. i Primenen., (1956) 289–319.
148. F.W. Stallard, Differential systems with interface conditions, Oak Ridge Nat. Lab. Rep. ORNL 1876, 1955.
149. F.W. Stallard, Functions of bounded variations as solutions of differential systems, Proc. Am. Math. Soc., 13 (1962) 366–373.
150. I. Stewart, The Lorentz attractor exists, Nature, 406 (2000) 948–949.
151. P. Thota, H. Dankowicz, Continuous and discontinuous grazing bifurcations in impacting oscillators, Phys. D, 214 (2006) 187–197.
152. J.D. Vasundara, A.S. Vatsala, Generalized quasilinearization for an impulsive differential equation with variable moments of impulse, Dynam. Syst. Appl., 12 (2003) 369–382.
153. A.S. Vatsala, J. Vasundara Devi, Generalized monotone technique for an impulsive differential equation with variable moments of impulse, Nonlinear Stud., 9 (2002) 319–330.
154. Th. Vogel, Théorie des systémes evolutifs, Gauthier-Villars, Paris, 1965.
155. F. Wang, C. Hao, L. Chen, Bifurcation and chaos in a Monod-Haldene type food chain chemostat with pulsed input and washout. Chaos, solitons and fractals, 32 (2007) 181–194.
156. G.S. Whiston, Global dynamics of a vibro-impacting linear oscillator, J. Sound Vibr., 118 (1987) 395–429.
157. S. Wiggins, Global Bifurcation and Chaos: Analytical Methods, Springer, New York, 1988.
158. L.A. Wood, K.P. Byrne, Analysis of a random repeated impact process, J. Sound Vib., 82 (1981) 329–345.

159. J. Yan, A. Zhao, J.J. Nieto, Existence and global activity of positive periodic solutions of periodic single-species impulsive Lotka-Volterra systems. Math. Comput. Model. 40 (2004) 509–518.
160. T. Yang, L.O. Chua, Impulsive control and synchronization of non-linear dynamical systems and application to secure communication, Int. J. Bifurcation Chaos Appl. Sci. Eng. (electronic resource), 7 (1997) 643–664.
161. Y. Zhang, J. Sun, Stability of impulsive delay differential equations with impulses at variable times, Dyn. Syst., 20 (2005) 323–331.
162. V.F. Zhuravlev, A method for analyzing vibration-impact systems by means of special functions, Mech. Solids, 11 (1976) 23–27.

Index